Nava Shaked, Ute Winter
Design of Multimodal Mobile Interfaces

Also of Interest

*Series: Speech Technology and Text Mining
in Medicine and Healthcare*
Amy Neustein (Ed.)
ISSN: 2329-5198

Published Titles in this series:

Speech and Language Technology for Language Disorders
Beals, Dahl, Fink, Linebarger; 2015
ISBN 978-1-61451-758-0, e-ISBN 978-1-61451-645-3

Speech and Automata in Health Care (2015)
Ed. by Amy Neustein
ISBN 978-1-61451-709-2, e-ISBN 978-1-61451-515-9

Text Mining of Web-Based Medical Content (2015)
Ed. by Amy Neustein
ISBN 978-1-61451-541-8, e-ISBN 978-1-61451-390-2

Computational Interpretation of Metaphoric Phrases
Weber Russell; 2015
ISBN 978-1-5015-1065-6, e-ISBN 978-1-5015-0217-0

Unimodal and Multimodal Biometric Data Indexing
Dey, Samanta; 2014
ISBN 978-1-61451-745-0, e-ISBN 978-1-61451-629-3

Design of Multimodal Mobile Interfaces

Edited by
Nava Shaked, Ute Winter

DE GRUYTER

Editors

Nava Shaked
HIT – Holon Institute of Technology
52 Golomb Street
POB 305 Holon 5810201
Israel
shakedn@hit.ac.il

Ute Winter
General Motors – Advanced Technical Center Israel
7 HaMada St.
7th Floor POB 12091
Herzliya Pituach 46733
Israel
ute.winter@gm.com

ISBN 978-1-5015-1084-7
e-ISBN (PDF) 978-1-5015-0273-6
e-ISBN (EPUB) 978-1-5015-0275-0
Set-ISBN 978-1-5015-0274-3

Library of Congress Cataloging-in-Publication Data
A CIP catalog record for this book has been applied for at the Library of Congress.

Bibliographic information published by the Deutsche Nationalbibliothek
The Deutsche Nationalbibliothek lists this publication in the Deutsche Nationalbibliografie;
detailed bibliographic data are available on the Internet at http://dnb.dnb.de.

© 2016 Walter de Gruyter Inc., Boston/Berlin
Cover image: "Circles" Original design by Brit Shaked
Typesetting: PTP Protago-TₑX-Production, Berlin
Printing and binding: CPI books GmbH, Leck
♾ Printed on acid-free paper
Printed in Germany

www.degruyter.com

Preface

This book is a collection of research articles and reviews in the area of Mobile Multimodality. It is looking at the field through various perspectives and areas of expertise in order to describe the main issues and challenges of putting together an optimized multimodal interface for mobile environments such as automotive, wearables, mobile applications, and avatars and virtual agents.

The book also raises usability questions that are of major concern to interface designers and architects in general, and even more so in a multimodal environment. A recurring usability theme is the importance of proper information or knowledge management. Today we are busy collecting and analyzing enormous amounts of data. Nevertheless, how much data is sufficient in order to create the "right" user experience? How do we present this information to the user? How can we introduce context into the multimodal paradigm, presenting information at the right time and in the right place?

What role do NLP engines play in the processing stage? How compatible are the current NLP engines with Mobile Multimodal systems? And how are NLP engines being standardized to support mobile multimodality?

And finally, does multimodality pose social, moral or behavioral questions that we need to answer while putting together all other components? In our highly technological, mobile and always-connected world where communication is essential and fast, and information is at everyone's fingertips from such an early age, how much freedom of choice and privacy should we support?

Acknowledgments

This book is the result of two years of continuous exploration of a field that is constantly changing and evolving. We wish to thank all the people who supported us in the process. First, we wish to thank the authors themselves, who agreed to share their research and vision – from academic work to real-time industry projects. We also want to thank HIT – Holon Institute of Technology and General Motors Advanced Technical Center – Israel for supporting us and allowing this book to develop. A big thanks to Dr. Laura Rosenbaun, our research assistant, who contributed greatly to the research and editing as well as to maintaining our morale. And finally to our book cover designer Brit Shaked, for sharing with us her creation.

We hope this book will inspire our students and colleagues in the fields of Mobility, Multimodality, and Interfaces, to keep developing and researching these exciting and continuously advancing domains.

What's in the book

Mobile Multimodality has touched many industries and is constantly developing and changing. The chapters in this book cover a wide variety of Mobile Multimodality issues. We shed light on central questions concerning the usability, concepts and complexity of Mobile Multimodality, taking into consideration not only some of the most interesting technological aspects but also the designers' and the users' concerns.

Most of the chapters deal with the challenge of deploying multimodal interface methods and technologies in a wide variety of contexts – mobile phone, automotive, contact center, wearables, etc. Typically, these systems collect contextual information for the purpose of personalization and optimization. Although the technology is becoming more reliable, the next challenge is dynamic adaptation to the requirements of the user, the tasks or the domain, as well as to environment and/or infrastructure constraints. One way to meet this challenge is to change the technology, while another approach is to gather large amounts of information about the user and the environment in order to create and present a different user interface in each context.

In the first chapter, *Introduction to the evolution of mobile multimodality*, Shaked and Winter lay the groundwork through explaining the basic terms and concepts of Mobile Multimodality and Human-Machine Interaction in general, and Multimodal Interaction in particular. The chapter explores the evolution of user behavior and needs as well as the development of new and exciting interaction technologies. The authors conclude by raising the "chicken and egg" question: What are the determining factors that influence new interface developments? Are they the users' new needs and requirements? Or are the latter only a response to the emergence of new innovative technologies?

Interaction is often about how the machine processes and perceives natural language automatically. In the chapter *Integrating natural language resources in mobile applications*, Dhal explores the contribution of Natural Language Processing (NLP) resources to the development of mobile applications with Multimodal Interfaces. The chapter describes the NLP components that are useful for integration into mobile applications. This integration, however, is extremely challenging and standardized specifications such as EMMA and MMI play a major role in reducing the effort required to create a seamless interface between NLP systems and other interfacing software.

Omnichannel architecture is an emerging technology that uses NLP components. In their chapter, *Omnichannel natural language – Bringing omnichannel and prediction to the multimodal user interface*, Brown, Walia and Urban discuss the fusion of contextual data for the purpose of prediction and service optimization. The authors describe the process of combining next generation UI technology with NLP, Omnichannel prediction, Big Data, and a real-time learning platform. Multimodal Interaction is defined and determined not only by the collection of data from various channels (SMS, voice calls, chats, web, etc.) but also by consolidation of the modalities used in those channels such as text, speech, gesture and more. The objective is to create a holistic contextual experience based on the user's current and past interaction history.

The utilization of various interaction data is also closely connected to the field of wearables. Both wearable computing and the Omnichannel infrastructure require building and maintaining an ecosystem to ensure effective interaction. The next chapter, *Wearable computing – The context challenge and a media ecology approach*, by Lawo, Logan and Pasher, explores the relationship between the user and her wearable environment. They argue that understanding the environment created by wearable computing as well as its psychological and social impact is crucial to implementing this technology. They introduce the concept of Wearable Ecology as a framework for understanding these impacts. The chapter reviews past and recent research on context frameworks for wearable computing and provides considerations for future directions in the development of wearable technology.

An additional use case of interaction and context data is laid out in *Spoken dialog systems adaptation for domains and for users* by Sun and Rudnicky. In this chapter Spoken Dialog Systems are the framework for a discussion of adaptation strategies that a machine can apply to accommodate user needs and domain specifics. How do individual users prefer to complete tasks? What is their mental model of the system and the domains that the system is designed for? How do user expectations and preferences, as well as the understanding of system capabilities and domain functionality, change over time? The authors significantly contribute to the questions raised through exploring adaptive strategies for language understanding and learning of user intention in applications that attempt to ensure a personalized user experience.

One of the most appealing interaction methodologies for mobile users are Avatars and Virtual Agents (A&VA). Virtual personal assistants, health caregivers, and game avatars have become part of our everyday activities. Their interaction modalities are discussed in the next chapter *The use of multimodality in avatars* and virtual agents by Shaked and Artelt. Like wearables, a special relationship between user and machine is created through A&VAs. Furthermore, the chapter argues that there is a direct relation between the development and progress of A&VA applications and the development and availability of multimodal interaction. The research question is how to determine the correct interaction design for A&VAs. The chapter examines a set of classification methods to characterize the types of user-machine relationship and interaction. This approach is then used to determine (a) which multimodal environment is best suited for the different interaction and relationship types, and (b) the optimal set of features for each case. As in the Omnichannel and the wearable environments, the authors recommend establishing a well-defined ecosystem in order to achieve the desired interaction.

Personal assistants may be presented to the user as voice persona, such as the popular Siri, Cortana, and Google Now speech-enabled personal assistants. A very compelling use case is the speech-enabling of an in-car infotainment system. The mobile industry has discovered this field of late, and has been developing tools to project user phone content and capabilities into the car infotainment system. For example, Siri can be used in a car via Apple CarPlay. To ensure efficient and pleasant communications of this type, user studies must be conducted in the specific driving context. Van Over, Molina-Markham, Lie, and Carbaugh have undertaken a near naturalistic study, which focusses on the cultural nature of communication between a human and machine in this setting. Managing interaction with an in-car infotainment system presents an investigation of turn exchange, a major aspect of any communication sequence. It further focusses on interactional misalignments as a result of misplaced verbal and multimodal cues. Which of those misalignments are culturally constituted? How do users adapt to such imperfect systems in order to accomplish their goals? Their findings lead to considerations for future multimodal in-vehicle system designs.

Two chapters discuss information organization on displays from different perspectives. Classification and organization of information: The case of the head up display by Heymann and Degani raises the question of how to organize and present information to drivers in a car environment using emerging technologies and features ranging from safety control to driver comfort to entertainment and communication capabilities. The chapter uses head up displays as an example of how overwhelming quantities of information can be categorized and prioritized. Their approach leads to a proposal of what elements of information should be presented to the driver on a head up display vs. a cluster or center stack display. Several existing head up displays are reviewed and a set of principles is presented to guide designers in their effort to organize information. The presented approach is relevant to display design in other environments and for other devices.

The second information organization chapter – *Towards objective methods in display design* by Shmueli-Friedland, Zelman, Degani and Asherov – discusses a complementary method for the display of information, in this case demonstrated through the example of the visual automotive instrument cluster. It applies a formal approach to the analysis and design of driver displays, using a series of methods involving domain experts' estimation as well as applying concepts of graph theory. The authors thus generate a cluster hierarchy that uncovers relationships among the informational elements. They illustrate the usefulness of their approach by analyzing a generic cluster display, and by revealing problems and missed opportunities in the design. In the end, they propose that this methodology can be generalized to include additional modalities, such as sound and haptics.

Contents

Kathy Brown, Anmol Walia, Prashant Joshi, Abir Chakraborty, and Sanjay Agarwal

Michael Lawo, Robert K. Logan, and Edna Pasher

List of contributing authors

Sanjay Agarwal
[24]7 Inc., Bangalore, India
sanjay.a@247-inc.com
Chapter 3

Detlev Artelt
AIXVOX GmbH
Germany
da@aixvox.net
Chapter 6

Ron Asherov
General Motors R&D
Advanced Technical Center Israel
rasherov@gmail.com
Chapter 9

Kathy Brown
[24]7 Inc., Campbell CA
kathy.brown@247-inc.com
Chapter 3

Donal Carbaugh
Department of Communication
University of Massachusettsm Amherst, MA
carbaugh@comm.umass.edu
Chapter 7

Abir Chakraborty
[24]7 Inc., Bangalore, India
abir.c@247-inc.com
Chapter 3

Deborah A. Dahl
Conversational Technologies, USA
dahl@conversational-technologies.com
Chapter 2

Asaf Degani
General Motors R&D
Advanced Technical Center Israel
asaf.degani@gm.com
Chapters 8 and 9

Michael Heymann
Department of Computer Science,
Technion, Israel Institute of Technology
heymann@cs.technion.ac.il
Chapter 8

Prashant Joshi
[24]7 Inc., Bangalore, India
prashant.joshi@247-inc.com
Chapter 3

Michael Lawo
Centre for Computing and Communication
Technologies (TZI) at the Universität Bremen
mlawo@tzi.de
Chapter 4

Sunny Lie
Pomona Department of Communication
California State Polytechnic University
slie@cpp.edu
Chapter 7

Robert K. Logan
Department of Physics and St. Michael's
College at the University of Toronto
logan@physics.utoronto.ca
Chapter 4

Elizabeth Molina-Markham
Independent Scholar
emolmark@gmail.com
Chapter 7

Brion van Over
Department of Communication
Manchester Community College
Manchester, CT
bvanover@manchestercc.edu
Chapter 7

Edna Pasher
EP Management Consultants
edna@pasher.co.il
Chapter 4

Alexander I. Rudnicky
Language Technologies Institute
Carnegie Mellon University
air@cs.cmu.edu
Chapter 5

Nava Shaked
HIT – Holon Institute of Technology
Israel
shakedn@hit.ac.il
Chapters 1 and 6

Yael Shmueli-Friedland
General Motors R&D
Advanced Technical Center Israel
yael.shmueli@gm.com
Chapter 9

Ming Sun
Language Technologies Institute
Carnegie Mellon University
mings@cs.cmu.edu
Chapter 5

Anmol Walia
[24]7 Inc., Campbell CA
anmol.walia@247-inc.com
Chapter 3

Ute Winter
General Motors R&D
Advanced Technical Center Israel
ute.winter@gm.com
Chapter 1

Ido Zelman
General Motors R&D
Advanced Technical Center Israel
ido.zelman@gm.com
Chapter 9

Nava Shaked and Ute Winter

1 Introduction to the evolution of Mobile Multimodality

Abstract: Human-Machine Interaction technologies and design have been at the center of both academic and industry research for several decades. Recent developments in the fields of mobility, sensors, Big Data, and their underlying algorithms have created an opportunity for a unique fusion of multiple factors to bring interaction to the next level. This chapter presents a model that describes the mutual influence and constant adaptation of technology and user needs in the area of Mobile Multimodality. We start by defining the key terms as they are used in the literature, with a focus on Mobility, User Interface, and Multimodality. We then describe major factors of influence such as social interaction, user behavior, contextual factors, and technological developments, among others. Having laid the groundwork for our model, we look closely at the two parties involved and sharing any interface: the users and the underlying technologies. We discuss both the evolution of user requirements and needs relative to mobile device interfaces as well as the historical development, readiness and availability of the underlying technologies. We then propose a cycle of mutual influence and adaptation between user needs and technology development, which leads to the constant evolution of Mobile Multimodal Interfaces.

1.1 User Interfaces: Does vision meet reality?

Movies and television shows traditionally transport us into a world of imagination and inspiration. They explore ideas and let us dream what could be if we just make it happen. One of the most appealing multimodal mobile interfaces of the 1980s was KITT, Michael Knight's car and communication partner in the series Knight Rider. Episode 12 in Season 2 contains the following spoken interaction sequence, when KITT had just failed to warn Michael in time of an imminent dangerous encounter with villains, and had to watch the incident from afar:

KITT: Michael! Michael! *(Pause while KITT waits to see if Michael got harmed)*
KITT: I'm terribly sorry, Michael, but he had such a head start.
Michael: It's ok, fell. What you'd get on the car?
KITT: It's out of scale of range. But I did run the plates. Unfortunately they were stolen from another car yesterday. Michael, are you alright?
Michael: Yeah, I am alright.

Apparently KITT has impressive conversational skills with a genuine human touch, makes observations via sensors with astonishing precision, and is connected to external networks and data. Its black displays though, representing the utterance only by some flashing red bars that simulate KITTs speech signal, seems highly unimaginative today after 30 years of continuous technology development. Still, KITT was undoubtedly multimodal and in many ways predicted the future.

After years of interfacing with KITT, what would Michael's reaction be to one of the current speech enabled multimodal interfaces? Despite the unquestionable progress in the technology and design of such interfaces, would he perceive this interaction to be natural and intuitive? Would he find the interface easy to use? Or perhaps users would prefer that multimodal interfaces be similar to KITT?

1.2 Discussion of terms: Mobility and User Interface

Like KITT, who accompanies Michael wherever he chooses to take his vehicle, multimodal user interfaces are often mobile. Mobility is therefore an important concept to examine in any discussion of multimodal interfaces.

1.2.1 Mobility

The Merriam-Webster dictionary defines mobility as being "able to move from one place to another", without or "with the use of vehicles", or as being "able to be moved". However, mobility should not be viewed only as a spatial phenomenon, associated with physical movement. As Ishii (2006, p. 347) points out, "mobility should be understood in a broader sense to include at least three interrelated dimensions of human interaction; namely, spatial, temporal, and contextual mobility" (see also Kakihara & Sorensen 2002). Users can interact with the interface and perform their intended task anywhere and anytime. Mobile interfaces thus have a huge impact on the social interaction between users and on the organization of daily tasks, thus contributing to social interaction and efficiency (Srivastava 2005; Schroeder 2010; Campbell & Kwak 2011). The third and more comprehensive factor, namely contextual mobility, may be of even higher social significance. Any human interaction, whether with humans or with devices, is context dependent (Hymes 1972; Gumperz 1982; Hagen et al. 2005; Tamminen et al. 2004; Stivers & Sidnell 2005). Space and time are only part of what constitutes context and contributes to the meaning of interaction; there are many other dimensions, such as cultural background, degree of familiarity among the interaction partners and the device, mood, use cases, etc. Mobile devices allow users to choose the context for their interaction, at least for some of the contextual dimensions such as, for example, choosing the degree of privacy.

Complementary, mobile devices should operate within their context to achieve effective interaction. "This context includes the network and computational infrastructure, the broader computational system, the application domain, and the physical environment" (Dix et al. 2000, p. 286). Although the first mobile device that comes to mind is the Smartphone, the list of mobile devices is long, each with its appropriate purposes: PDAs, wearable computers, portable media players, game consoles, cameras, pagers, PNDs, tablets, and more. The device design must include the infrastructure and architecture to adapt to the purpose of the device, the variety of use cases that a user can perform, the desired style of interaction, and the typical context for such use cases. In addition, the design must take into account connectivity to networks and external servers; the interaction sequences for the use cases and domains, including the integration of contextual data; and the choice of input and output modalities. All of this must be dealt with within the physical constraints of mobile interfaces, which are often small and with limited embedded computing power.

1.2.2 User Interface

This discussion leads to a closer look at user interfaces for mobile environments. Often the first issue to come to mind regarding user interfaces is design: design of displays or controls, design aesthetics, etc. Some user interface definitions make this their focus, such as Jacob, who states that, "a user interface is that portion of an interactive computer system that communicates with the user. Design of the user interface includes any aspect of the system that is visible to the user" (2004, p. 1821). Another exemplary definition focuses on the software aspect of an interface saying that it is "the software component of an application that translates a user action into one or more requests for application functionality, and that provides to the user feedback about the consequences of his or her action." (Myers & Rosson 1992, p. 196). Galitz mentions that "the user interface is the part of a computer and its software that people can see, hear, touch, talk to, or otherwise understand or direct. The user interface has essentially two components: input and output" (2007, p. 4). In other words, they emphasize the user experience of interaction, of providing input to and receiving output from a system. More user interface definitions or descriptions can be found in Dix et al. (2004) and Jacko (2012), to mention just a few.

The plurality of definitions attests to the complexity of the challenge. If the purpose of the user interface is to enable meaningful interaction between a user and a system, perhaps we need to broaden our scope and understand what is involved in human-machine interaction. According to Galitz (2007, p. 4) "human-computer interaction is the study, planning, and design of how people and computers work together so that a person's needs are satisfied in the most effective way. HCI designers must consider a variety of factors: what people want and expect, what physical limitations and abilities people possess, how their perceptual and information processing systems work, and what people find enjoyable and attractive. Designers must also consider technical characteristics and limitations of the computer hardware and software." We adopt this perspective and understand the fundamental challenge as aligning knowledge about the user's expectations and needs with the mode of operation and features of the machine, to create an interface that facilitates the desired interaction between both sides.

It is worthwhile to examine at all mentioned partners in the interaction starting with the machine or system. In the context of multimodal mobile interfaces, the machine mostly involves hard- and software – not only in close proximity to the user but also connected, for instance, to Internet servers or cloud-based services and applications. Many mobile software algorithms rely on statistical processes that are limited in their predictive power and accuracy, yet demand high computation power for fast processing, such as speech or gesture recognition algorithms. These are only two examples of the bandwidth of limitations and constraints of such systems and technologies. It also demonstrates how complex the machine or system can be. Moreover, this complexity is constantly growing due to the increasing amounts of data to learn from, new applications, improving and new technologies, and ways to combine them all. When designing the user interface, the interaction must be optimized for reducing the inherent complexity of the machines to simple and easy to perform interaction sequences, providing error prevention and recovery strategies while avoiding user confusion.

The user, on the other hand, with his physical and cognitive capabilities, personal tastes, wishes and expectations, does not easily change beliefs and behaviors. Although he may be willing to explore new approaches, he will also most likely have a firm point-of-view on how the interaction should – or should not – work. The challenge lies in understanding the wide range of user constraints, tastes, expectations and preferences, as well as the underlying cultural norms that shape user expectations and behavior. Any of this is also influenced by the situation and context in which the interaction takes place. The user may be focusing solely on the interaction with the system or may be occupied primarily with another task such as driving a car. The user may very well have expectations about the system's knowledge in dependency of the context.

1.2.3 User-centered design

Given the complexity of the systems and the variety of users, it is not surprising that "the design of effective interfaces turns out to be a surprisingly difficult challenge, a fact that is attested to by the often-frustrating experiences we all encounter in today's digital world" (Salvendy 2012, p. 1179). Over the last two decades new approaches have attempted to overcome interface design ineffectiveness. One is the paradigm of user-centered design (Nielsen 1993), a multidisciplinary design approach that relies on the active involvement of users to improve the understanding of user and interface requirements. With the interface being iteratively designed, developed and evaluated in user and usability studies, user-centered design is widely regarded as overcoming the limitations of traditional system-centered design (Mao et al. 2005). During the design phase user persona are developed based on user studies, and during the development phase usability testing is conducted based on pre-defined test scenarios and measures for the evaluation of user satisfaction (Nielsen 1993; Norman & Draper 1986; Vredenburg & Butler 1996). Classic user-centered design assumes an existing set of technologies that has to be configured to a user group. But due to the rapid and dynamic evolution of technologies, the approach needs to respond flexibly to *ad hoc* technology integration and prospective user expectations.

1.2.4 Teamwork

The user-centered design approach should not be misunderstood as giving the user entire control over the system. Another concept that has been explored and found useful in interface design is the concept of teamwork, in which the user and the system form a collaborative team to achieve the interaction goal (Grosz & Kraus 1996; Horvitz 1999; Rich, Sidner & Lesh 2001; Klein et al. 2004). Teamwork can be defined and implemented in a user interface through more than one paradigm. The user and the system can be partners and peers, transferring back and forth control and initiative within an interaction sequence according to a shared decision-making process. Alternatively, control and decision-making can be shifted to one of the partners according to the demands of the task at hand. Goldman and Degani (2012) define a list of seven attributes necessary to form a team and discuss different computational models that can be applied in the teamwork approach to human-machine interaction. The attributes are: mutual commitment; models for understanding (beliefs, intentions, history, states, and actions); common language; coordination; responsiveness; mutual awareness resulting in trust, humor and elegance of interaction; and accomplishment.

1.2.5 Context

Indirectly present in these attributes is awareness of the situation in which an interaction occurs. During spoken communication, such as a human using a speech system, we understand, interpret and judge any dialog in its communication situation, i.e. its context (Hymes 1972). Context dependency requires that we consequently need to design the interface for the users' preferred interaction practices and styles for each of the typical interface situations and use cases (Carbaugh et al. 2013). The same is true for other interface modalities. Contextual factors can also influence the kind of user requests from a mobile interface, such as heavy traffic may lead to the interface's recommendation of a different route than the most recommended.

Furthermore "effective interface design ... requires a deep appreciation of the work domain itself: interface design strategies that are appropriate for one category of domains may very well not be appropriate for another" (Salvendy 2012, p. 1179). User studies have shown that users who are commuting or doing errands mostly prefer an efficient interaction style with sequences optimized for interaction time, while users who are gaming are more focused on pleasantness and entertainment (Winter, Shmueli & Grost 2013).

1.3 System interaction: Moving to Multimodality

1.3.1 User input and system output

How will the interactions between user inputs and system outputs optimally be conducted? The widely used term "nonverbal communication" implies that humans express themselves and interact with their environment by a variety of multimodal cues, without the explicit use of words. Any kind of "human interaction with the world is inherently multimodal" (Turk 2014, p. 189). We send multimodal cues through touch, gestures, gaze, facial expressions, distance, and physical appearance, among others. We also communicate nonverbally through the use of voice by paralanguage, including rate, pitch, volume, style, as well as prosodic features such as rhythm, intonation, and stress. We are also endowed with multiple senses to accept and process external information from a variety of modalities. Research has proven that humans hardly distinguish between humans and machines in their interactions (Reeves & Nass 1996). Consequently, it is not surprising that "as in human-human communication ... effective communication is likely to take place when different input devices are used in combination" (Sebe 2009, p. 19).

1.3.2 Multimodality

The understanding of the term multimodality in the context of interaction is less diverse than the understanding of the term user interface itself. Previously a multi-modal system has been defined as a system that has the capacity to communicate with a user through different types of communication modes and to extract and convey meaning automatically (Nigay & Coutaz 1993).

Similarly, Oviatt (2012) starts an overview on multimodal interfaces by stating that "multimodal systems process two or more combined user input modes – such as speech, pen, touch, manual gestures, gaze, and head and body movements – in a coordinated manner with multimedia system output". Oviatt et al. (2000) discussed the usefulness and benefits of multimodal interfaces including, among others: greater input and output flexibility; efficient switching of modes to achieve shortcuts in the interaction; more alternatives and naturalness in the interaction according to user preferences; and enhanced error prevention and recovery. These examples show that carefully designed multimodal interfaces are superior to uni-modal. There is a variety of literature available that provides overviews on multimodality, such as Oviatt 2012; Oviatt & Cohen 2015; and Turk 2014.

From the perspective of technology, recent decades have witnessed incredible advances in many fields of hard- and software that recognize the user's multimodal cues to a system, such as speech recognition, gesture recognition, gaze recognition, facial expression and emotion detection, and touch interfaces, to name just a few. With recognition accuracy still often dependent on machine learning algorithms and on vast amounts of reliable data, these technologies have not yet achieved their full potential. However, they are advanced enough to demonstrate their power and natural suitability for interface design within the context of the ubiquitous mobile computing environment. Likewise, system output modalities have evolved along the lines of visual and auditory cues, such as sounds, speech, haptics, and visual screens.

With the advancement of multimodal systems and the pervasiveness of mobility, systems are now offering more and more sophisticated ways for using the data collected in the interaction, and using statistical and machine learning heuristics to analyze and optimize the interaction experience. Contextual Multimodality is a natural by-product of this approach.

1.3.3 Combining modalities

In addition to further progress regarding the accuracy and quality of the technologies themselves, the grand challenge in multimodal interface design is to combine modalities in a meaningful way and according to the natural preferences of the target user group, the relevant use cases, and domain and context of the interface. Modality fusion refers to the integration or combination of two or more modalities into one interaction step. Two terms have emerged to describe natural multimodal fusion solutions that originate from the observed multimodal behavior of humans. Bolt's "Put That There" system (Bolt 1980), combining voice and gesture, is considered the pioneering demonstration of fused modalities. The other term is Transmodality, referring to putting in the foreground of the user interface the preferred or most efficient modality at each interaction step and context, and thus providing an effective user experience within and across modalities rather than providing all modalities in parallel throughout the interaction sequence.

Ideally, the interface should allow the user to switch between modalities or to create modality fusions throughout the interaction in her preferred way, with the interface itself outputting each interaction step in one or more modalities according to the user's preferences. This type of interface would give users a more holistic experience when interacting with the machine. The question is: What are the assessment criteria for the choice of modalities? When would the user think that her mobile interface is simple, efficient, intuitive, pleasant or even exciting?

1.4 Mobile Multimodality: The evolution

1.4.1 Technology compliance to user needs

Perhaps the biggest conceptual leap that Steve Jobs made in the early days of Apple was to recognize that high-tech devices not only could, but must be friendly. In addition he said, "Design is not just what it looks like and feels like. Design is how it works." (Walker 2003). Jobs understood that design simplicity should be linked to making products easy to use.

In light of the incredibly fast pace of technology development as well as the constant adaptation of users to their own preferences and the environment, today we can go one step further and claim that *Mobile Multimodality uses state of the art technologies to create innovative user interfaces for human-machine interaction that comply with constantly changing user and market needs*. We will further substantiate this statement in this section.

This interrelationship between technology and user needs poses a great challenge for any Mobile Multimodality design effort: in spite of growing technological complexity, the interface has to be intuitive and natural to its users. To meet this challenge Mobile Multimodality is a multidisciplinary industry comprised of technologies from the fields of Computing, Applied Mathematics, Engineering, Psychology, Artificial Intelligence and, of course, UIX Design.

It is interesting to look back at the historical outline of interaction technology development, in order to better understand the current status of multimodal user experience. In accordance with our statement above, two key factors have determined the development and shape of today's Mobile Multimodal User Interaction: technology readiness and availability, and user requirements and needs.

1.4.2 Technology readiness and availability

The last three decades of development in three major technological fields – Computing, Internet and Cellular Mobile – were essential for the emergence of multimodal user interfaces. With every advance in chip design and processing power, memory capacity, data management and storage, and cloud computing, user interaction designers could experiment with yet another form of Human-Machine Interaction. Meisel (2013), in describing the evolution of software, notes that increasing processing power, as well as the prevalence of the Internet and cloud computing, created a massive demand for Software as a Service. A major impact on software evolution, according to Meisel, is the ubiquity of information. Data availability anywhere, coupled with improved search technologies, laid the groundwork for improved usability of human-computer interfaces.

Moreover, the mobile infrastructure, characterized by its connectivity and accessibility features, enabled a high level of interaction using multiple sensor inputs, large databases and multimedia processing. Mobile products have become more powerful and operating systems more generic, allowing hardware and software to work together to create a plethora of multimodal applications. The relationship between multimodality and smartphones is nicely described by Oviatt and Cohen (2015), who claim that the proliferation of multimodal interfaces with their sophisticated capabilities has been driven by the rise of the smartphone. "On a global scale, this trend toward expanded development of multimodal interfaces is accelerating with the adaption of smart phone and other mobile devices in developing regions." (p. xxi)

The mobile world is still struggling with security and privacy issues as well as with energy and power challenges but, overall, the multimodal user experience is becoming more holistic in nature. We are no longer looking at each component separately but at the prospects of combining them to work together. Let's go one step further to explore the readiness and availability of the multimodal technology.

1.4.3 The readiness of multimodal technology

Stanciulescu (2006) claims that, in communication, "mode" refers to the communication channel used by the two entities that interact. Based on implied sensorial systems as well as on motor systems, he talks about gesture, vocal, graphical, tactile as modes. Each communication mode has an associated interaction type that characterizes it. But a mode can have more than one interaction type. For example: the Tactile mode can be reflected in both text and touch interaction types. The Gesture mode can be associated with hand, face and lip movement as well as eye tracking.

It is interesting to note that while human interaction modes are finite and limited by human capabilities, due to technological development and processing engines machines are capable of implementing virtually infinite interaction types. With the huge number of sensors in our mobile devices, which are connected to other data sources (such as in the cloud), mobile multimodality is now accessible, available and easy to use. Moreover, with the development of input and output technologies, the machine is able to use many modes to communicate with the human, creating a more natural interaction.

Oviatt and Cohen (2015, p. 136–137) present a table with a summary of commercial applications that support multimodal interaction, where the distinguishing categories are: *application type* (automotive, wearable, education, etc.), *platform* (PC, cell phone, wearable, etc.) and *manner of interaction* for which they list the possible human modes for input modalities. In Dumas et al. (2009) we find the notion of "action" and "perception" to describe the interaction loop. This creates another layered look at the process.

We propose to include not only human mode inputs but also other sensor and data input, resulting in four main categories: *mode, interaction type, processing engine, media sensor or device*. Some examples are shown in Fig. 1.1:

- *Mode:* the type of "mode" used in the classical definition of Stanciulescu (2006).
- *Interaction type:* how this mode is realized as a human input. Each communication mode has an associated interaction type that characterizes it. However, a mode can have more than one interaction type. For example: the Tactile mode can be reflected in both text and touch interaction types. The Gesture mode can be associated with hand, face and lip movement as well as eye tracking.
- *Processing engine:* which technology is being used to process the interaction type.
- *Media sensor or device:* the apparatus, usually part of the hardware or software infrastructure, enabling the interaction to be realized.

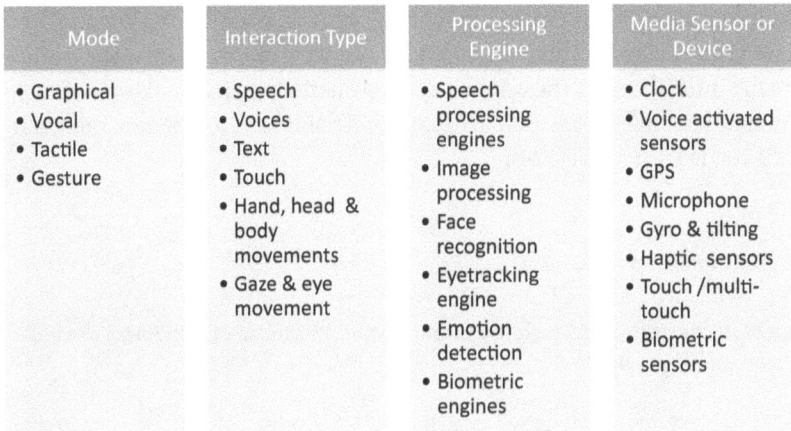

Mode	Interaction Type	Processing Engine	Media Sensor or Device

• Graphical
• Vocal
• Tactile
• Gesture
	• Speech
	• Voices
	• Text
	• Touch
	• Hand, head &
	  body
	  movements
	• Gaze & eye
	  movement
		• Speech
		  processing
		  engines
		• Image
		  processing
		• Face
		  recognition
		• Eyetracking
		  engine
		• Emotion
		  detection
		• Biometric
		  engines
			• Clock
			• Voice activated
			  sensors
			• GPS
			• Microphone
			• Gyro & tilting
			• Haptic sensors
			• Touch /multi-
			  touch
			• Biometric
			  sensors

Fig. 1.1: Pillars of Mobile Multimodality with some of their types.

With these four pillars we can make the distinction between the different layers of interaction on the one hand and the technology on the other.

With the continuous development of input/output technologies coupled with Big Data analysis and algorithms in areas such as machine learning, data mining, and artificial intelligence, among others, as well as new and better processing engines and sensors, Mobile Multimodality is becoming an available reality for every device and user.

1.4.4 User requirements and needs

The second factor greatly influencing human-machine interaction is constant change in user requirements, needs and habits. What role do users play during the innovation of a new technology? Bogers et al. (2010) ask whether users take part in the innovation process and, if they do, why and for what reasons. Furthermore, they examine how users contribute deeply to the evaluation of new ideas and innovations. Von Hippel et al. (2011) claim that users themselves can be the innovators, not just helpers to producers who innovate. Bogers et al. (2010) add that the rapid pace of technological change, globalization and increasing user sophistication mean that more and more users will have more and more opportunities to innovate or contribute to producer innovations.

There is an unwritten contract between users and technology inventors to keep upgrading and creating more options for users to "fulfill fantasies". A user interaction technology is introduced and adopted on a wide scale. It changes the way we communicate with our PC, mobile device, our car, or our TV game. The new technology can be part of an application, a service, or a tool; often one cannot separate the interface from the product itself. Nonetheless, it is making a change in our human-machine interaction to exclude old habits and adopt new, more natural and effective communication.

1.4.5 Cycle of mutual influence

We propose to describe these changes in a cyclic model of change and development such as in Fig. 1.2.

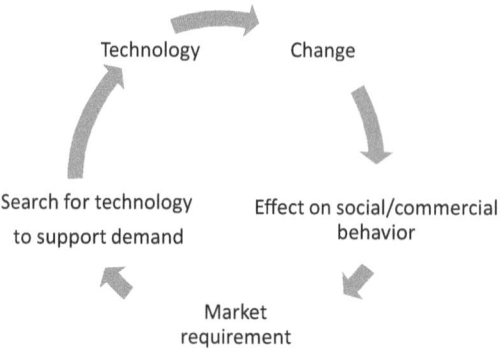

Fig. 1.2: The cyclic nature of mutual influence between technology and user.

This cycle presents the relationship between an emerging new technology and its adaptation and consumption by users as it creates a change in their habits and behaviors, leading them to require and demand new features. Thus, for example, one can talk about the new habits and behavioral changes created by smart mobile user interfaces starting with unimodal and bi-modal input technologies and ending with full multimodal designed products at the demand of users at an accelerating rate of acceptance. Our model claims that user interaction technology is directly affected by user expectations, and that users play a major role in the innovation cycle.

For example, let's go back to Apple's sensational announcement on January 9, 2007 regarding one of many features of the iPhone that changed our mobile behavior forever – the multi-touch user interface. We can trace touch screen technology all the way back to IBM building the first touch screens in the late 1960s, and, in 1972, Control Data released the PLATO IV computer, a terminal used for educational purposes that employed single-touch points in a 16 × 16 array as its user interface. Substantial research on multi-touch technology was done in the early 1980s in the University of Toronto's Input Research Group, Bell Labs, and other R&D/academic institutions. However, between 1999 and 2005 a company named Fingerworks developed various multi-touch technologies that were acquired by Apple. With the launch of the iPhone in 2007, mainstream users experimented with multi-touch technology, adopting it very quickly and requesting additional functions and utilities. Although Apple did not invent multi-touch, it did invent the multi-touch user interface for mobile phones and this interface technology gained enormous popularity, leaving behind companies that did not foresee the effects of this technological change (Multi-touch, https://en.wikipedia.org/wiki/Multi-touch, 2016).

Another technology that has shaped the development of Mobile Multimodal interfaces is the appearance of mobile personal assistants (PAs). Speech recognition technology has been around for many decades in one variation or another, but it was not able to penetrate the masses until the arrival of Siri. This personal assistant, used mostly through speech interaction, was able to do what no other speech application did before – create an awareness of the possibilities of multimodal interaction. Siri, followed by Google Voice, Cortana, Echo, Ivee, and many others, have infiltrated everyday life and are creating a demand for many more task types and flexibility in interaction.

According to Olsson (2014) user expectations are a factor affecting the actual user experience design in human-computer interaction. There are different layers of expectations constituting a framework that is providing an understanding of the spectrum of user expectations from technologies. Surveying users about their expectations, values, needs, and requirements from emerging technologies can shed light on which innovations may succeed and how people conceptualize new technologies. This insightful information is relevant for the development and design of products and services, and for the user interaction experience.

Olsson (2014) talks about four layers of expectations, which we would like to reference in the context of Mobile Multimodality technology:

1. **Desirers:** Expectations of what users would like to have as added value. These expectations can be the foundation for new applications and future technologies. For example: using data collection and analysis to optimize multimodal systems and using Machine Learning methods to predict usage and personalize accordingly.

2. **Experience based assumptions:** Expectations based on past experience of users' assumptions and expectations of a product/technology/service. For example: which modalities of interaction will apply in which situation? In mul-

timodal design for wearables, there is an experience-based expectation that interaction will remain private and protected yet easy to use.

3. **Society and societal norms:** Expectations based on current norms and trends that are acceptable by society or by other users, including criteria such as usability, aesthetics, and social acceptability.
   For example: Accessibility issues are a societal concern and multimodal interaction and interfaces are expected to support them.

4. **Must have expectations:** Expectations for minimal requirements for user acceptance, with the emphasis on negative experience and taking into consideration unacceptable features or the absence of "must have" ones. For example it is expected that Multimodal Interaction design will provide options for I/O technologies so that the user can choose depending on the circumstances. Noisy environments are not suitable for speech input while texting is unacceptable in moving vehicles.

The main objective of this chapter is to establish the connection between technology readiness and user needs and requirements. We claim that the interaction technologies that underlie user interfaces are greatly dependent on and influenced by user expectations and experience. Understanding Olsson's four layers of user and societal expectations is crucial to the success of interface design in general, and is even more important in complex Mobile Multimodal interfaces where multiple factors can influence the user experience. To create a multimodal holistic experience that meets user needs and expectations we must first methodically understand all the layers of user expectations.

1.5 Conclusion

We have shown how user interaction theories have evolved over decades creating a new interaction paradigm, which is comprised of interface design, users and interaction technologies. We have described how Multimodality is a key to meeting the requirements and needs of today's users, with Multimodality benefitting from advances in mobile technologies and underlying infrastructures such as the Internet, computing environment, and data collection and usage. We have established the connection and mutual influence between the evolution of technology innovation and changing user expectations. We have shown that many multimodal interaction technologies have become user favorites due to the combination of existing computing technologies with mobile devices and sensors. In these ways we have demonstrated and established our initial statement that *Mobile Multimodality uses state of the art technologies to create innovative user interfaces for human-machine interaction that comply with constantly changing user and market needs.*

Meisel (2013) claims that we will benefit from technology's evolution if we manage to control it and provide maximum benefits to society. He suggests that "a key result of this advancing technology is tighter connection between human intelligence and computer intelligence" (p. xvii). We believe that this view is particularly relevant to our discussion. Multimodality is being pushed to new levels because we can – because it is no longer unimaginable to think about the fusion of so many levels of computing, communication, data and applications. But Multimodality is also being driven by the ongoing study of user preferences, priorities, suggestions, and demands for more intuitive and easier interfaces. This is the "tighter connection" that we seek to encourage.

We are in the midst of a groundbreaking era, with the digital revolution, mobility and multimodal technologies playing central roles. The driving forces that we describe in this chapter guarantee the unstoppable forward momentum of this transformation.

Abbreviations

PDA   Personal Digital Assistant
PND   Personal Navigation Device
PC   Personal Computer
HCI   Human Computer Interaction
UIX   User Interface XML (Extensible Markup Language)
ASR   Automatic Speech Recognition
TTS   Text To Speech
GPS   Global Positioning System
R&D   Research & Development
PA   Personal Assistant
I/O   Input Output

References

Bogers, M, Afuah, A & Bastian, B 2010, 'Users as innovators: a review, critique, and future research directions', *Journal of Management*, vol. 36, no. 4, pp. 857–875.

Bolt, RA 1980, '"Put-that-there": Voice and gesture at the graphics interface', *ACM Comput. Graphic*, vol. 14, no. 3, pp. 262–270.

Campbell, SW, Kwak, N 2011, 'Mobile communication and civil society: Linking patterns and places of use to engagement with others in public', *Human Communication Research*, vol. 37, no. 2, pp. 207–222.

Carbaugh, D 2005, *Cultures in conversation*, Routledge, London, New York.

Carbaugh, D,Winter, U, Van Over, B, Molina-Markham, E & Lie, S 2013, 'Cultural analyses of in-car communication', *Journal of Applied Communication Research*, vol. 41, no. 2, pp. 195–201.

Dix, A, Rodden, T, Davies, N, Trevor, J, Friday, A & Palfreyman, K 2000, 'Exploiting space and location as a design framework for interactive mobile systems', *ACM Transactions on Computer-Human Interaction (TOCHI)*, vol. 7, no. 3, pp. 285–321.

Dumas B, Lalanne, D & Oviatt S 2009, 'Multimodal Interfaces: A survey of principles, models and frameworks', in *Human Machine Interaction*, LNCS 5440, eds D Lalanne and J Kohlas, Springer-Verlag, Berlin, pp. 3–26.

Galitz, WO 2007, *The essential guide to user interface design: An introduction to GUI design principles and techniques,* Wiley Publishing, Inc., Indianapolis, IN.

Goldman, CV & Degani, A 2012, 'A team-oriented framework for human-automation interaction: Implication for the design of an advanced cruise control system', *Proceedings of the Human Factors and Ergonomics Society Annual Meeting*, Boston, MA: Human Factors Society, October 22–26, SAGE Publications, pp. 2354–2358.

Grosz, BJ & Kraus, S 1996, 'Collaborative plans for complex group action', *Artificial Intelligence*, vol. 86, no. 2, pp. 269–357.

Gumperz, John J 1982, *Discourse strategies*, Cambridge: Cambridge University Press.

Hagen, P, Robertson, T, Kan, M & Sadler, K 2005, 'Emerging research methods for understanding mobile technology use', *Proceedings of the 17th Australia Conference on Computer-Human Interaction: Citizens Online: Considerations for Today and the Future*, Computer-Human Interaction Special Interest Group (CHISIG) of Australia, p. 1.

Horvitz, E 1999, 'Principles of mixed-initiative user interfaces', *Proceedings of the SIGCHI Conference on Human Factors in Computing Systems*, ACM CHI, pp. 159–166.

Hymes, D 1972, 'Models in the interaction of language and social life', in *Directions in sociolinguistics: The ethnography of communication*, eds J Gumperz & D Hymes, Blackwell, New York, NY, pp. 5–71.

Ishii, K 2006, 'Implications of mobility: The uses of personal communication media in everyday life', *Journal of Communication*, vol. 56, no. 2, pp. 346–365.

Jacob, RJ 2004, 'User Interfaces', in *Encyclopedia of Computer Science*, 4th edn, eds A Ralston, ED Reilly & D Hemmendinger, John Wiley & Sons, Chichester.

Kakihara, M & Sorensen, C 2002, 'Mobility: An extended perspective', *System Sciences, 2002. HICSS. Proceedings of the 35th Annual Hawaii International Conference on IEEE*, pp. 1756.

Klein, G, Woods, DD, Bradshaw, JM, Hoffman, RR & Feltovich, PJ 2004, 'Ten challenges for making automation a "Team Player" in joint human-agent activity', *IEEE Intelligent Systems*, vol. 19, no. 6, pp. 91–95.

Mao, J, Vredenburg, K, Smith, PW & Carey, T 2005, 'The state of user-centered design practice', *Communications of the ACM*, vol. 48, no. 3, pp. 105–109.

Meisel, W 2013, *The software society: Cultural and economic*, Trafford Publications, Bloomington, Indiana.

Myers, BA & Rosson, MB 1992, 'Survey on user interface programming', *Proceedings of the SIGCHI Conference on Human factors in Computing Systems*, ACM, p. 195.

Multi-touch 2016, (Wikipedia article). Available from: https://en.wikipedia.org/wiki/Multi-touch [14 January 2016].

Nielsen, J 1993, *Usability engineering*, Morgan Kaufmann Publishers Inc., San Francisco, CA.

Nigay, L & Coutaz, J 1993, 'A design space for multimodal systems: concurrent processing and data fusion', *Proceedings of the INTERACT'93 and CHI'93 Conference on Human Factors in Computing Systems*, ACM, p. 172.

Norman, DA & Draper, SW 1986, *User centered system design; New perspectives on human-computer interaction*, Lawrence Erlbaum Associates, Inc., Mahwah, NJ.

Olsson, T 2014, 'Layers of user expectations of future technologies: an early framework', *CHI'14 Extended Abstracts on Human Factors in Computing Systems*, ACM, pp. 1957–1962.

Oviatt, S 2012, 'Multimodal interfaces', in A. Sears & J. Jacko (eds.), *The human-computer interaction handbook: Fundamentals, evolving technologies and emerging applications* (3rd edn), Erlbaum: Mahwah, NJ, pp. 405–430.

Oviatt, S, Cohen, P, Wu, L, Duncan, L, Suhm, B, Bers, J, Holzman, T, Winograd, T, Landay, J & Larson, J 2000, 'Designing the user interface for multimodal speech and pen-based gesture applications: state-of-the-art systems and future research directions', *Human-Computer Interaction*, vol. 15, no. 4, pp. 263–322.

Oviatt, S, Cohen, P, 2015, *The Paradigm Shift to Multimodality in Contemporary Computer Interfaces*, Morgan and Claypool, San Rafael, CA.

Reeves, B & Nass, C 1996, *The media equation: How people treat computers, television, and new media like real people and places*, Stanford, CA, CSLI Publications and Cambridge University Press.

Rich, C, Sidner, CL & Lesh, N 2001, 'Collagen: Applying collaborative discourse theory to human-computer interaction', *AI Magazine*, vol. 22, no. 4, pp. 15–25.

Salvendy, G 2012, *Handbook of human factors and ergonomics* (4th edn), John Wiley & Sons, Hoboken, New Jersey.

Schroeder, R, 2010, 'Mobile phones and the inexorable advance of multimodal connectedness', *New Media & Society*, vol. 12, pp. 75–90.

Sebe, N 2009, 'Multimodal interfaces: Challenges and perspectives', *Journal of Ambient Intelligence and Smart Environments*, vol. 1, no. 1, pp. 23–30.

Sidnell, J & Stivers, T 2005, Multimodal interaction. Special issue of *Semiotica*, 156

Srivastava, L 2005, 'Mobile phones and the evaluation of social behaviour', *Behaviour & Information Technology*, vol. 24, issue 2, pp. 111–129.

Stanciulescu, A 2006, *A transformational approach for developing multimodal web user interfaces*. Dissertation, Studies in Management, Catholic University De Louvain, School of Management, Belgian Laboratory of Computer Human Interaction.

Tamminen, S, Oulasvirta, A, Toiskallio, K & Kankainen, A 2004, 'Understanding mobile contexts', *Personal and Ubiquitous Computing*, vol. 8, no. 2, pp. 135–143.

Turk, M 2014, 'Multimodal interaction: A review', *Pattern Recognition Letters*, vol. 36, pp. 189–195.

von Hippel, EA, Ogawa, S & Jong, J 2011, 'The age of the consumer-innovator', *MIT Sloan Management Review*, vol. 53, no. 1, pp. 27–35.

Vredenburg, K & Butler, M 1996, 'Current practice and future directions in user-centered design', *Usability Professionals' Association Fifth Annual Conference*, Copper Mountain, Colorado.

Walker, R 2003, 'The guts of the machine', *New York Times Magazine*, November 30. Available from: http://www.nytimes.com/2003/11/30/magazine/30IPOD.html [14 January 2016].

Wickens, C, Lee J, Liu Y & Gordon Becker S 2004, *An Introduction to Human Factors Engineering* (2nd edn), Pearson Prentice Hall, Upper Saddle River, NJ.

Winter, U, Shmueli, Y & Grost, TJ 2013, 'Interaction styles in use of automotive interfaces', in *Proceedings of the Afeka AVIOS 2013 Speech Processing Conference*, Tel Aviv, Israel.

Deborah A. Dahl

2 Integrating natural language resources in mobile applications

Abstract: This paper discusses integrating natural language technology in mobile applications. It begins by discussing why natural language understanding is a valuable component of mobile applications. It then reviews some tasks that can be done with natural language understanding technology and gives an overview of some natural language systems that are currently available for integrating into applications. Because natural language understanding is a dynamic field, and the systems discussed may become out of date, the paper also discusses general criteria for selecting natural language technology for use in mobile applications. Despite the wide availability of natural language resources, the task of adding natural language understanding functionality to applications is made more difficult because most of the APIs to natural language systems are entirely proprietary. The paper emphasizes how standards can greatly simplify the task of integrating natural language resources into applications and concludes with an overview of two standards from the World Wide Web Consortium. Extensible Multimodal Annotation (EMMA) and the Multimodal Architecture and Interfaces (MMI) specification reduce the amount of effort required to integrate natural language understanding into other applications by providing a uniform interface between natural language understanding systems and other software.

2.1 Natural language understanding and multimodal applications

2.1.1 How natural language improves usability in multimodal applications

As smartphone and tablet applications increase in features and sophistication, it becomes more and more difficult to successfully navigate their user interfaces. Because of this rich functionality, typical graphical user interfaces must include multiple menus, buttons and gesture actions that users must navigate to find the features that they want to access. This means that users often have to explore many paths before they can reach their goal, frequently backtracking when they've gone down the wrong path. In addition, terminology both across and within applications is frequently inconsistent (clicking on "settings", "options", "preferences", or "account" all can lead to similar functionality in different applications). This situation results from the fact that screen space is extremely limited on mobile devices, so that less functionality can be exposed at the level of a main screen. Instead features must be nested in submenus and sub-submenus. If the submenus and sub-submenus are organized into categories with clear semantics, users have a chance at finding what they need, but frequently the semantics is not clear. As even smaller devices (such as smart watches) become available, with even smaller screens, this problem worsens. In addition, current trends toward flat design contribute to the problem by hiding user interface options until a specific, non-intuitive, touch gesture is performed.

All of this means that learning a new application of even moderate complexity on a mobile device is often time-consuming and frustrating. Direct searching for functions, rather than traversing menus, of course, is another user interface option, and many applications support search in addition to menu interaction. However, open ended search with keyboard input is very painful on a mobile device. To avoid the problem of typing on tiny keyboards, current smartphones include very capable speech recognition, so simply entering keyword search terms has become much less difficult. However, voice input alone is not the answer, because simple keyword searches often lead to many unwanted results, burying the user's real target in a long list of irrelevant suggestions. This leaves users in the position of trying different keywords over and over to try to find search terms that will get them the results they need, without ever knowing whether the application even has the capability they're looking for. The problem is even more severe for users who are less familiar with mobile applications than the general population or for users who have cognitive disabilities.

This problem is not only evident with mobile devices. As the Internet of Things expands, it will become more and more difficult to use conventional graphical interfaces to interact with an environment that includes possibly hundreds of connected objects, each with its own capabilities.

More natural interfaces are needed – interfaces that support natural forms of interaction and which don't require mastering a new user interface for each application. Spoken natural language can provide a uniform interface across applications by allowing users to state their requests directly, without navigating through nested menus or trying to guess the right keywords. Natural language also reduces the need for the user to learn special input idioms like application-specific swiping gestures.

Using spoken natural language is much simpler and more convenient than current touch-based approaches. With natural language, users speak their requests in a natural way. For example, in a shopping application, users could say things like "I'm looking for a short-sleeved cotton woman's sweater" instead of navigating through a set of menus like "women's → tops → sweaters →

short-sleeved → cotton". A user of a tourist information app could say things like "What's the best bus to take from here to the Museum of Art?" instead of "transportation → local → bus → destinations → cultural → Museum of Art". Similarly, a natural language tourist information app could also use translation services to answer questions like "how would I ask for directions to the Art Museum in French?" These kinds of use cases provide a strong argument for natural language as a routine mobile user interface.

2.1.2 How multimodality improves the usability of natural language interfaces

Not only is there a strong case to be made that natural language improves the usability of multimodal applications, but there is also a powerful benefit of multimodality in improving the effectiveness of natural language interfaces. The graphical modality especially can complement natural language interaction in multiple ways. The simplest benefit is probably just a display of speech recognition or speech understanding results. If the system has made an error, it should be clear from the display of the results. The application could also display alternative results that the user could select from in case of a mistake. This is much faster and more efficient than error correction with a voice-only dialog. As another example, the graphical modality can guide the user toward context-appropriate utterances through the use of graphics that set the context of the expected user response. This can be as simple as displaying "breadcrumbs" on the screen to indicate the dialog path to the current topic of discussion. Speech can also be combined with digital ink and pointing gestures to support natural multimodal inputs such as "Tell me more about this one", or "Do you have that in red?" Finally, system output displayed as text can speed the interaction by bypassing the need to speak lengthy system responses out loud.

2.2 Why natural language isn't ubiquitous already

There are two major reasons why, despite these benefits, natural language isn't already a ubiquitous user interface. The first is that spoken natural language understanding and speech recognition have not until recently been perceived as sufficiently reliable to support a wide variety of types of user interaction. However, this is rapidly changing, and the technologies are now mature enough to support many different types of applications. Unfortunately, while natural language understanding is becoming more common in mobile apps, this capability is not always available to developers because it is built in to the platform and lacks APIs.

Some well-known systems that include natural language understanding components, such as Apple Siri, Google Now, Microsoft Cortana, or Amazon Echo, are essentially closed systems, without APIs, or with only limited APIs, that developers could leverage in their own applications. This is very limiting for the goal of natural language understanding as a ubiquitous interface, because it would be impossible for these individual vendors to ever integrate all the potential natural language applications to their current closed systems. Instead, for natural language to be truly ubiquitous developers will need to be able to independently create natural language UIs for any application.

A second, perhaps less obvious, reason that spoken natural language is not more common is that easy-to-use APIs and integration tools have not been available. This problem is being partially addressed as natural language technology is increasingly made available through the RESTful APIs (Fielding & Taylor 2000) that are becoming more and more common for accessing many different types of web services. While RESTful APIs do make integration easier, they don't fully solve the problem because it is still necessary to code to specific natural language processing resource's APIs. Standards such as Extensible Multimodal Interaction (EMMA) (Dahl 2010; Johnston et al. 2009a, Johnston 2016, Johnston et al. 2015) and the Multimodal Architecture (Barnett et al. 2012; Dahl 2013b), as discussed below, will also play a role in reducing development effort by providing more uniform APIs to natural language processing functionality.

However, despite the fact that standards are only beginning to be adopted, there nevertheless are an increasing number of web services and open source systems which are currently available for integration into applications. Full natural language understanding, as well as specific natural language tasks such as part of speech tagging, parsing, stemming, translation, and named entity recognition are only a few of the natural language processing technologies that are available. Because of this, developers who would like to use natural language processing in applications have many more resources available to them than they would have had only a few years ago.

2.3 An overview of technologies related to natural language understanding

Before we discuss the details of different natural language processing resources, it is useful to situate natural language understanding in the context of some important related technologies. The relationship between natural language processing and these other technologies addresses the question of which other technologies would be needed to build a specific system. It also clarifies the architecture of products where natural language understanding is bundled with one or more related technology. In practice, natural language understanding is nearly always used in conjunction with one or more of these other related technologies.

Figure 2.1 illustrates the relationship between some of these technologies and natural language understanding (in the gray box). Natural language understanding analyzes text, which can originate from a variety of sources, including (1) existing documents (from the Internet or from databases) or (2) a user's typed text or (3) recognized speech creating a representation in a structured format. Once natural language understanding has taken place, an interactive dialog manager can act directly on the user's request and/or the results can be stored for later use, for example, to support applications such as search.

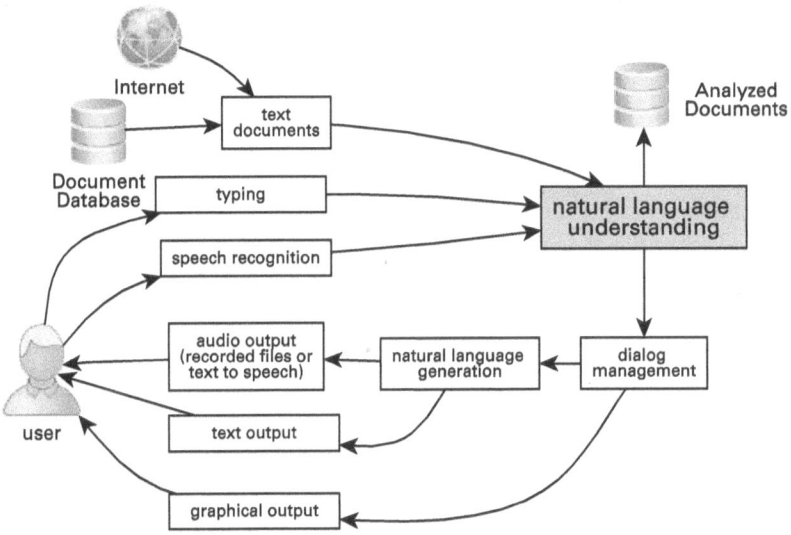

Fig. 2.1: System configurations with natural language understanding.

This paper will review some currently available natural language processing resources, describe how they can be used in mobile applications, and discuss how they can be used with W3C standards to provide the basis of a new generation of natural, easy to use, mobile user interfaces. The specific software resources detailed here are changing rapidly, so any discussion of current systems is likely to become out of date in the near future. In order to extend the value of this paper beyond the lifetime of the specific systems discussed here, we also discuss general categories of systems and how to select software for a specific application.

2.4 Natural language processing tasks

The goal of natural language processing is to create structured representations from unstructured natural language, the kind of language that we use to communicate with other people. Within this general characterization of natural language processing, there are many different types of structures that may be produced, depending on the goals of the application. Some systems produce several types of structures, often in a pipelined fashion where one task passes its result to the next task in the pipeline, each task adding new levels of structure.

There are many currently available natural language understanding systems, but there is no single system that can meet the natural language processing requirements for every application. Thus, it is important to carefully consider what needs to be accomplished in order to select the right system for any specific application. We begin by reviewing a few of the most common natural language processing tasks and their use cases.

Classification

One of the simplest natural language processing tasks is classification. This is a characterization of natural language inputs into categories, or *intents*. This is also called *text classification*. For some applications, this may be all that is needed for very useful results. Even if the users' speech is very verbose, indirect and full of hesitations and false starts, classification can be very effective because the systems are developed or *trained* on just this kind of data. Table 2.1 shows some results from a classification system that classifies natural language utterances into different emotions (Dahl 2015).

One important application of classification is sentiment analysis, where sentiments expressed in text are represented. This is very useful, for example, in automatic processing of product reviews. Product reviews can be automatically classified as presenting positive, negative or neutral opinions about a product as well as the strength of the sentiment expressed. Similarly, classification has been used to classify tweets (Lee et al. 2011) into various categories of interest.

Tab. 2.1: Emotion classification from text.

Input	Classification
I'm feeling unhappy right now	sad
This is gross	disgusted
I had a frightening experience	afraid
I'm in a cheerful mood today	happy

Named entity recognition

Named entity recognition is a natural language processing task that attempts to find references to specific individuals or organizations in text. A simple example of named entity recognition would be, given the sentence, "President Obama spoke to Congress about the Supreme Court", identifying Obama as a person, and identifying Congress and the Supreme Court as organizations. This would be useful for topic spotting applications – for example, if a company wants to find out if their products are mentioned in blogs.

Interactive systems

Interactive systems are a large category of natural language applications that involve a back and forth dialog with a user; consequently, the specific natural language task in interactive systems is to structure the user's input in such a way that the system can act on it and provide a response to the user.

One type of interactive system is *question answering*. The goal of question answering is to answer questions about content, either generic knowledge such as Wolfram|Alpha (Claburn 2009), IBM Watson (Ferruci et al. 2010) or knowledge about specific documents (for example, product manuals) such as NLULite (NLULite 2015).

Another type of interactive system where natural language processing is used is *spoken dialog systems*. Spoken dialog systems are more general than question-answering systems, because they engage in many different types of dialogs, not just question answering (although they often do answer questions). Dialog systems include *dialog managers* (as shown in Fig. 2.1) that make decisions about what to say next in a conversation, taking into account the earlier parts of the conversation and previous interactions with the user. The more sophisticated a dialog manager is, the better it will be at adapting its behavior taking into account information about the context.

Currently, one important type of interactive dialog system is the personal assistant. Applications of this type include systems such as Siri, Google Now, Amazon Alexa, or Cortana. Personal assistants allow users to do tasks such as get information, manage their schedules, and interact with connected objects in the Internet of Things. As an example, natural language processing in a personal assistant application could convert something like "I need to add three heads of lettuce to my shopping list" to a structured format like the one shown in Fig. 2.2. A newer type of personal assistant is the Enterprise Assistant that represents a company and assists customers with tasks such as voice shopping or product support. Enterprise assistants are based on similar technologies as personal assistants. Examples of these enterprise assistants include Nuance Nina (Nuance Communications 2015) and Openstream Eva (Openstream 2015).

Traditional voice-only Interactive Voice Response (IVR) systems, as used in call centers, are also a type of interactive system. They typically help callers with self-service tasks such as checking on an order, reporting an electrical outage, checking bank balances. Interactions with IVRs tend to be relatively restricted and inflexible compared to personal assistant applications because they focus on a specific task.

Interactive applications in call centers also frequently use classification technology, discussed above, for *call routing* applications. The ideas behind these systems are based on the original work of Gorin, Riccardi and Wright (Gorin et al. 1997). Natural language call routing systems allow callers to express their concerns in ordinary language, with prompts like "In a few words, tell me what you're calling about today." The classification system then assigns the caller's query to a specific category, which allows the call to be directed to an agent with knowledge of that category.

Classification extracts a fairly coarse meaning; a more fine-grained meaning can be obtained from *entity extraction*, or finding important key-value pairs in a user's utterance, in addition to finding the overall intent of an utterance, as shown in Fig. 2.2. In Fig. 2.2, the overall classification of the user's intent is "add item to shopping list". There are three key-value pairs that have been extracted from the utterance – shopping_item: iceberg lettuce, measure: heads, and number: 3. Systems of this type are trained on examples of utterances expressing the intents and entities of interest to the application; other information in the utterance is ignored. Thus, the training process is extremely important. Poorly designed training data can easily result in a system that is unusable.

```
shopping_item: iceberg lettuce
measure: heads
number: 3
```

Fig. 2.2: Structured format for "I need to add three heads of lettuce to my shopping list".

Natural language technology in spoken dialog systems is usually based on entity extraction with key-value pairs and may also include text classification for recognizing overall intents.

Linguistic analysis

Some natural language systems have the goal of analyzing language, independently of a specific application task. *Lexical analysis*, one type of linguistic analysis, means analyzing words; for example, *stemming*, or finding the roots of words, and *lexical lookup*, finding words in a dictionary. Tasks that analyze the grammatical structure of inputs include *part of speech tagging*, or assigning parts of speech to words in an input, and *parsing*, or analyzing the input's syntactic structure. These tasks can be used standalone in applications such as grammar checking and correction, but they can also be used as intermediate tasks for applications that are aimed at finding the structured meaning of an utterance. As an example, Tab. 2.2 shows the part of speech tagging results for "I had a frightening but very interesting experience".

Tab. 2.2: Part of speech tags (Penn Treebank tagset) for "I had a frightening but very interesting experience".

Word	Part of Speech Tag	Meaning of Tag
I	PRP	Personal pronoun
had	VBD	Past tense verb
a	DT	Article
frightening	JJ	Adjective
but	CC	Coordinating conjunction
very	RB	Adverb
interesting	JJ	Adjective
experience	NN	Common noun

Syntactic analysis, as shown in Fig. 2.3, is a detailed analysis of the grammatical relationships in a sentence, showing how the words are related to each other syntactically. Syntactic information, for example, includes modification relationships, such as the fact that "frightening" and "interesting" describe or modify "experience". This type of analysis could be useful on its own in grammar-checking applications, in foreign language learning, or it can be part of the process of creating the fuller meaning analysis shown in Fig. 2.2.

For more details about natural language processing technology and applications, Dahl (2013a) provides a high-level overview of natural language processing and Jurafsky and Martin (2008) provide an in-depth technical resource for speech and language technologies.

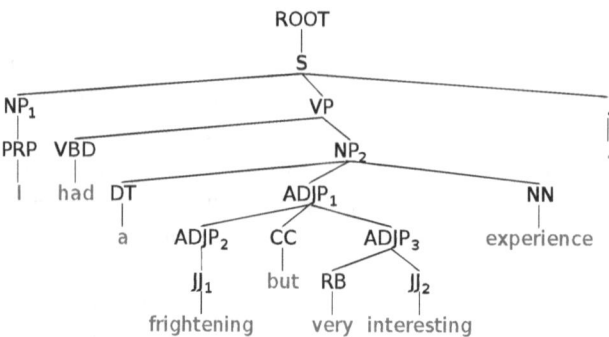

Fig. 2.3: Syntactic parse tree for "I had a frightening but very interesting experience".

2.4.1 Accessing natural language technology: Cloud or client?

Now that we have discussed the different types of tasks that can be accomplished with natural language understanding, we can turn to the question of how to access natural language processing from an application.

Web services

The most common style of API for online natural language processing systems is RESTful Web Services, with results provided in JSON or XML. This is an increasingly common format for Web Service results, and is applicable to many web-based APIs, not only those used for natural language. Clients for this kind of API can include native smartphone or tablet interfaces such as iOS, Android, or Windows, as well as more generic browser-based HTML5/JavaScript-based applications. In a RESTful approach, the client, for example a web browser, sends a request to a URL (the endpoint) with text or speech to be processed, along with possible application-specific parameters, usually using HTTP GET or POST, and receives a result with the results of the natural language processing. In most cases the details of the request and result formats are vendor-specific, but later on we will see examples of RESTful natural language APIs using the W3C EMMA (Johnston et al. 2009b) and Multimodal Architecture (Barnett et al. 2012) standard Life Cycle Events.

For commercial applications, the vendors of natural language web services will also usually support on-premise hosting, so that a commercial application does not have to be dependent on the vendor for mission-critical services.

Server-based software

Another alternative to accessing natural language systems is software hosted on an internal server. This will be necessary for commercial applications of most open source software, since open source projects don't usually have the resources to maintain web services which get any significant traffic. As noted above, a hosted or on-premise solution is also appropriate for mission-critical systems. Of course, a RESTful interface can also be used internally within an enterprise for natural language processing services.

Natural language software on the user's device

This is the only model for accessing interactive natural language processing which does not require the user to be online, and there are obvious advantages to this. Software on the device will work whether or not the cloud is accessible. At this time, there seem to be very few natural language systems that are available for installation locally on users' devices (although there have been some interesting experiments – see (Kelly 2015) for example). However, it is theoretically possible for Java-based systems like Stanford Core NLP (Stanford Natural Language Processing Group 2014) or OpenNLP (OpenNLP 2015) to run on Android devices, keeping in mind that they would have significant storage and processing requirements. It would be interesting if future systems were to explore this approach.

2.4.2 Existing natural language systems

This section summarizes some existing natural language systems, current as of this writing. We start with some general considerations.

Research and commercial systems

Open source research systems such as the Stanford CoreNLP tools (Stanford Natural Language Processing Group 2014), OpenNLP (OpenNLP 2015) and NLTK (Loper & Bird 2002) are especially suitable for learning about natural language processing. There is no support for production applications, although there are message boards that can offer community support. These research systems are also among the few natural language tools that can provide structural analyses like part of speech tagging or syntactic analysis. An advantage of open source systems is that it is also possible for developers to modify them; for example, to add a new language.

Cost

Costs to get started with any natural language understanding are generally low. Open source systems like Stanford CoreNLP, OpenNLP and NLTK are free to use, although licensing terms may make some commercial applications difficult. Almost all commercial services have a low-volume free tier for development or research, but the licenses for production can become expensive at high volumes. It's important to be aware of the full range of pricing if the application will be used commercially.

Development/training needed or ready to use?

Application-specific interactive dialog systems such as Wit.ai (wit.ai 2015), Microsoft LUIS (Microsoft 2015), or api.ai (api.ai 2015) require developers to create application-specific entities and intents that describe the objects that will be referred to in the application. This can be done manually or it can be done programmatically if a lot of data has to be entered. Wit.ai and api.ai both provide web interfaces and APIs that developers can use to supply training data programmatically for their application. Systems that perform linguistic analysis (like Stanford CoreNLP) will normally not require developers to train the system itself for specific applications because those systems are application independent; however, it is necessary to add an application component to these systems that maps the generic meanings to application-specific meanings. They also require training for new languages. If one of these systems has to be extended to a new language then it will be necessary to supply it with training data, usually through the use of an annotated corpus. Developing a new language is a very technical and time-consuming task that would be difficult for most developers.

Bundled with other technology

Some natural language technology is bundled with other technologies (common examples of bundled technology are information extraction from text or dialog management) and can't be used independently of the other components. This is a benefit if the other technologies have value for the intended application, but if they are not needed in the application, then the bundling may make it impossible to use the natural language technology for other purposes.

2.4.3 Natural language processing systems

Here we list a number of current systems. The following is by no means a complete list of available natural language understanding systems, but it represents a representative snapshot of the options available as of this writing. New and updated natural language processing systems become available frequently, so this list is not definitive.

api.ai

api.ai (api.ai 2015) is a natural language processing system that accepts text or speech input for use in interactive systems. It has an online developers' interface, a REST interface and native SDKs for iOS, Android, Cordova, and HTML. Right now only English is supported, but major European and Asian languages are planned.

Watson

The IBM Watson Developer Cloud (IBM 2015) supports question answering based on a corpus of documents for use in interactive systems. It also has a structural analyzer, "Relationship Extraction" that finds semantic relationships between components of sentences. In addition, the IBM Watson suite has related services such as speech to text and object recognition. These services are currently in beta.

LingPipe

LingPipe (alias-i 2015) is a set of natural language processing tools from Alias-I with a Java API. The tasks it supports are sentence segmentation, part of speech tagging, named entity extraction, coreference resolution, and Chinese word segmentation. Except for Chinese word segmentation, only English is supported.

Linguasys

Linguasys (LinguaSys 2015) offers two APIs. The Carabao Linguistic Virtual Machine™ can extract named entities, perform sentiment analysis, and perform foreign language search. It includes a RESTful interface for interactive systems. Linguasys supports 23 languages.

LUIS (Microsoft Project Oxford)

LUIS (Microsoft 2015) is an application-specific REST API for natural language processing. It returns JSON that includes the entities and intents found in the user's utterance.

OpenEphyra

OpenEphyra (OpenEphyra 2015) is an open source, question answering system in Java developed at Carnegie-Mellon University.

OpenNLP

OpenNLP (OpenNLP 2015) is an open source Java system which supports a number of linguistic analysis tasks such as part of speech tagging and parsing. OpenNLP supports English and Spanish.

NLTK

NLTK (nltk.org 2015) is a set of open source natural language processing tools in Python that includes part of speech tagging and parsing. The license does not permit commercial use.

NLULite

NLULite (NLULite 2015) processes texts and then allows users to ask questions about those texts. NLULite has a client-server architecture. The server runs only on Linux x86_64 or Mac OS X. The Python client is open source but the server is proprietary. NLULite only supports English.

Pandorabots

Pandorabots (Pandorabots 2015) is a keyword based natural language processing system bundled with question answering and dialog management based on Artificial Intelligence Markup Language (AIML). A RESTful interface is available. Pandorabots is programmed using AIML (Wallace 2014).

Stanford CoreNLP

Stanford CoreNLP (Stanford Natural Language Processing Group 2014) is an extensive set of open source Java programs that support linguistic analysis as well as named entity recognition in English, Spanish, or Chinese.

Watson Developer Cloud (IBM)

The Watson Developer Cloud offers several services that perform natural language processing functions. Question and Answer is an application for question answering in the healthcare and travel domains. There is also a Natural Language Classifier for classifying texts into categories and related services such as machine translation and speech to text.

Wit.ai (Facebook)

Wit.ai (wit.ai 2015) provides natural language processing based on a RESTful web service. The developers' interface allows developers to train a system (called an "instance") with their own examples. The trained system can then be accessed via HTTP to process written or spoken inputs. The inputs are analyzed into *intents*, or overall classifications, and *entities*, which are specific key-value pairs within the intent. For example, for the analysis in Fig. 2.2, the overall intent might be "shopping list", and the result would also include three entities; "measure", "shopping_item" and "quantity". Wit.ai also includes a wide range of generic, commonly used entities for concepts like numbers, money, times and dates. It is free to use since its acquisition by Facebook. Wit.ai currently supports English as well as nine European languages.

Wolfram|Alpha

Wolfram|Alpha (Claburn 2009) has an API which can be queried to get answers to a wide of types of questions in multiple subject areas, focusing on science, technology, geography and people. Wolfram Alpha also has tools for building custom natural language applications using their natural language technology. Wolfram|Alpha is only available for English.

Table 2.3 summarizes the above discussion.

Tab. 2.3: Summary of natural language systems.

System	Task	Open Source?	Access	Cost	Languages
Api.ai	Classification and entity extraction, optional speech input	No	REST API	Usage-based tiers	11 languages
IBM Watson	Question answering, entity extraction	No	REST API	In beta, free for now	English
LingPipe	Linguistic analysis	No	Local install	Tiered, includes a free tier	English?
Linguasys	Interactive dialogs, linguistic analysis, classification	No	REST API	Call for quote	16 languages
Microsoft (Luis)	Interactive dialogs	No	REST API	Currently free (beta)	English, others unknown
OpenEphyra	Question answering	Yes, Java	Application-specific	Free	English
OpenNLP	Linguistic analysis	Yes, Java	Application-specific	Free	English, Spanish
NLTK	Linguistic analysis, classification	Yes, Python	Application-specific	Free, but no commercial applications allowed	17 languages
NLULite	Information extraction/ question answering	Client is open source (Python), server is closed source	Server and client	Free for intro version, 300 euros for company license	English
Pandorabots	Interactive dialogs	No	REST API	Usage tiers based on volume	English
Stanford CoreNLP	Linguistic analysis, classification, named entity recognition	Yes, Java	Application-specific	Free for research, contact for commercial use	English, Spanish, Chinese, French, German, Arabic
Wit.ai	Classification and entity extraction, optional speech input	No	REST API	Free	Major European languages
Wolfram|Alpha	Question answering	No	REST API	Free for development, contact for commercial use	English

2.4.4 Selection Criteria

What should developers look for when selecting natural language processing technology for an application? Here are some technical criteria to consider.

1. Does the software support the intended application?
2. For commercial systems, the licensing has to be consistent with the intended use. Can the system be used commercially?
3. How much application-specific development is required and is it necessary for the development to be done by natural language processing experts?
4. Is the software accurate enough to support the intended application? No natural language understanding software will always be completely accurate for all inputs, so occasional errors are to be expected, but they must not be so frequent that the application is impossible.
5. It is important to consider latency in interactive systems. When a user is actively interacting with systems, the latency between the user's input and the system's response must be minimized. For offline systems that process existing text, for example analysis of online product reviews, latency is much less of an issue.
6. Are the required languages supported by the software and how well are they supported? Most NLU systems support English, and if other languages are supported, English is usually the language with the most complete implementation. If other languages are needed in an application, it is important to make sure that the implementations of the other languages are of high enough quality to support the application. If the required languages are not supported, how difficult and costly will it be to add them?
7. Is the system actively maintained and is development continuing? This is particularly important for closed source systems because critical software errors may make it impossible to keep using the system if the system is not actively being maintained. However, it is important to be aware of this concern even for open source systems. Theoretically, anyone can repair problems with open source software, but it may not always be possible to find developers with the required expertise. A related issue is the stability of the vendor. If the software has been developed by only one or two key people, what happens if the key people become unavailable?

2.5 Standards

As we can see from the discussion above, there is currently a rich variety of natural language processing software available for incorporation into mobile applications. The available software has many capabilities and covers many languages. However, it can still be difficult to make use of this software. Nearly every natural language web service or open source system has its own, vendor-specific, API. This means that developers who want to change the service they use, combine information from multiple services, or support several languages that use different services will have to learn a different API for every system. A standard interface to natural language understanding systems would make it much easier to support these use cases. This section discusses two standards developed by the World Wide Web Consortium; one for representing natural language results, and another, more general, standard for communication within multimodal systems, which will greatly improve the interoperability of natural language understanding technologies.

2.5.1 EMMA

The details of existing APIs for natural language results are currently largely proprietary; however, they nearly always contain the same basic information – key-value pairs, confidences, alternative results and timestamps, for example. The proprietary APIs differ primarily only in formatting. Thus, there's no reason to have multiple different formats for representing the same information, and this common information can be standardized. We discuss two standards here.

EMMA (Extensible Multimodal Annotation) (Johnston et al. 2009a, Johnston et al. 2015), a specification published by the World Wide Web Consortium, provides a standard way to represent user inputs and their associated metadata in any modality, including but not limited to speech, typing, handwriting, and touch/pointing. EMMA 1.0 documents include an extensive set of metadata about the user input, as well as an application-specific representation of the meaning of the input. The current official standard is EMMA 1.0, but a new version, EMMA 2.0, is under development. Some ideas which have been discussed for EMMA 2.0 include the ability to represent system outputs and the ability to return incremental results while the user input is still in progress. Figure 2.4 is an example of an EMMA document where the task is emotion identification from text.

While we will not go into the details of each annotation individually, it is worth pointing out several of the more important annotations, shown in bold in Fig. 2.4. The *interpretation* ("<emma:interpretation>") is an EMMA element repre-

```
<emma:emma
  xmlns:emma="http://www.w3.org/2003/04/emma"
  xmlns:emotion="http://www.w3.org/2009/10/emotionml"
  xmlns:xsi="http://www.w3.org/2001/XMLSchema-instance" version="1.1"
      xsi:schemaLocation="http://www.w3.org/2003/04/emma
http://www.w3.org/TR/2009/REC-emma-20090210/emma.xsd">
      <emma:one-of emma:tokens="i am happy yet just a little afraid at the same
time" id="oneof13">
        <emma:interpretation emma:confidence="0.18257418583505536"
id="interp59">
          <emma:derived-from composite="false" resource="#initial1"/>
          <emotion category-set="http://www.w3.org/TR/emotion-
voc/xml#everyday-categories">
            <category confidence="0.18257418583505536" name="afraid"/>
          </emotion>
        </emma:interpretation>
        <emma:interpretation emma:confidence="0.15075567228888181"
id="interp60">
          <emma:derived-from composite="false" resource="#initial1"/>
          <emotion category-set="http://www.w3.org/TR/emotion-
voc/xml#everyday-categories">
            <category confidence="0.15075567228888181" name="happy"/>
          </emotion>
        </emma:interpretation>
      </emma:one-of>
    </emma:emma>
```

Fig. 2.4: EMMA representation for "I am happy yet just a little afraid at the same time".

senting the application-specific meaning of the input, based on the original user input ("emma:tokens"). In this case the input was "I am happy yet just a little afraid at the same time". There are actually two possible meanings – "happy" and "afraid", since the utterance mentioned two emotions. These are ordered according to the confidence the emotion interpreter placed on each result, enclosed in an "emma:one-of" tag.

The information inside the <emma:interpretation> tag is application-specific and is not standardized. The decision not to standardize the application semantics was motivated by the desire to maximize flexibility in representing semantics, which is an active research area, and can vary greatly across applications. In the case of Fig. 2.4 the application-specific semantic information is represented in Emotion Markup Language (EmotionML) (Schröder et al. 2009), a W3C standard for representing emotion. Figure 2.4 is a fairly complete EMMA document, but many of the annotations are optional, and a complete EMMA document can be much smaller, if a simpler document meets the application's requirements. In fact, in some use cases it is useful to create two EMMA documents for the same input; a simple one which can be sent to a mobile client with limited resources and used directly in an interactive dialog system, and a fuller document, including much more metadata about the utterance, which could be stored on a server as proposed in EMMA 2.0. The information in the fuller document could be used for detailed logging and analysis, which are very important for enterprise-quality deployments.

2.5.2 MMI Architecture and Interfaces

The second standard we will discuss here is the Multimodal Architecture and Interfaces (MMI Architecture) standard (Barnett et al. 2012, Dahl 2013b, Barnett 2016). The W3C Multimodal Architecture and Interfaces standard provides a general communication protocol for controlling multimodal components, including natural language processing systems. This is useful for the task of integrating natural language into applications because it defines a specific, standard, API between the natural language processing components and the rest of the system. It also supports adding different, complementary, modalities, such as speech recognition or handwriting recognition. Since the API is standard, it doesn't have to be changed if a new natural language processing technology is added to the system.

The MMI Architecture is organized around an Interaction Manager (IM), which coordinates the information received from user inputs to Modality Components (MCs). MCs are capable of processing information from one or more modalities, such as speech recognition, handwriting recognition, touchscreen, typing, object recognition or other forms of user input. Figure 2.5 shows an example of an MMI Architecture-based system with an IM coordinating an interaction that includes six modalities.

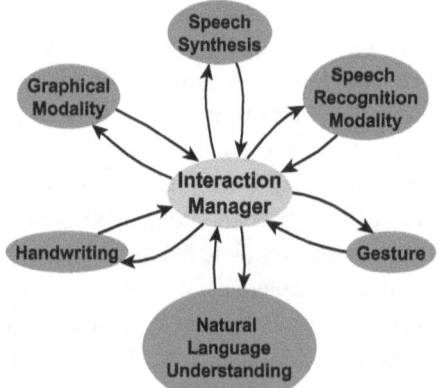

Fig. 2.5: An MMI Architecture-based system with six modalities.

The MCs are black boxes with respect to each other and to the IM; all communication between IMs and MCs takes place through a set of Life-Cycle events. The Life-Cycle events are very generic and are widely applicable across applications. They include events for starting, canceling, pausing and resuming processing, as well as events for returning the results of processing and other data. EMMA-formatted data is used in the Data field of Life-Cycle events that represent user inputs. Figure 2.6 is an example of a "DoneNotification" Life-Cycle event, whose purpose is to return the results of processing (Dahl 2015). In this case, the result (not shown in detail) is an EMMA document, the result of natural language processing on a user input. In addition to the EMMA-formatted result, the DoneNotification event also includes metadata that identifies the context that the message is related to ("mmi:Context"), whether or not the processing was successful ("mmi:Status") and the source and target of the message ("mmi:Source" and "mmi:Target"). More information about the MMI Architecture can be found in the standard itself (Barnett et al. 2012), or a summary of the standard (Dahl 2013b, Barnett 2016).

```
<mmi:mmi
   xmlns:mmi="mmi">
   <mmi:DoneNotification
       mmi:Context="nlClient0957"
       mmi:RequestID="requestID1559"
       mmi:Source="ctNLServer"
       mmi:Status="success"
       mmi:Target="ctNLClient">
   <mmi:Data>
     <emma:emma
(EMMA result)
     </emma:emma>
   </mmi:Data>
   </mmi:DoneNotification>
</mmi:mmi>
```

Fig. 2.6: DoneNotification MMI Life-Cycle event with EMMA result.

The availability of standards formats such as EMMA and the MMI Architecture will make it much more convenient to use natural language processing technology in many types of applications by greatly enhancing interoperability across different natural language understanding systems.

2.6 Future directions

While natural language processing technology has greatly advanced in the last few years, there is still a lot of work to be done. The application-specific natural language processing systems discussed in this chapter are very good at handling simple commands and recognizing basic intents that involve only a few entities. Specifically, they are very capable at extracting simple meanings from utterances specific to their domains.

The application-independent systems, on the other hand, are good at analyzing syntax and some generic semantics, such as recognizing named entities. However, two important capabilities are missing. First of all, application-independent systems attempt to extract very little application-independent meaning. Consequently, a specific application that is based on an application-independent system is required to build its own meaning analysis system. Depending on the system, this can be accomplished either by providing examples of utterances and their meanings, or by providing rules, depending on whether the system is statistical or rule based. This leads to an application development bottleneck, even for systems that only need to understand simple commands. Work on semantic role labeling (Gildea & Jurafsky 2000) may help close the gap between application-independent systems and real-world applications, but there is still a gap between semantic-role labeled data and application concepts that must be bridged by the developer.

The second problem is that semantically complex utterances are very difficult for today's application-specific systems to handle. Semantically complex utterances include, for example, those with a negative meaning, utterances with multiple commands, or utterances with time references. For example, utterances including negatives might include something like "Please record every Phillies game unless they're playing the Nationals", said to a DVR. Even an apparently simple utterance with multiple commands like "Put 2 bottles of orange juice and 1 bag of baby carrots on my shopping list" would be very hard to handle because simpler systems are not very good at associating the correct numbers with the items mentioned. Although currently users are more familiar with interacting with systems with simpler language, I believe these types of capabilities will be increase in importance as users become accustomed to interacting with technology by voice.

2.7 Summary

This chapter has discussed a wide variety of types of natural language processing resources that are available for developers who wish to incorporate natural
language processing into mobile applications. We also discussed specific systems
and selection criteria for finding the right natural language processing technology
for particular applications. Finally, we presented a brief overview of two relevant
standards, EMMA and the MMI Architecture, that show significant promise for
accelerating the integration of language understanding into mobile applications.

Abbreviations

EMMA   Extensible Multimodal Interaction
REST   Representational State Transfer
API   Application Programmer's Interface
MMI   Multimodal Architecture and Interfaces
W3C   World Wide Web Consortium
IVR   Interactive Voice Response
JSON   Javascript Object Notation
XML   Extensible Markup Language
URL   Uniform Resource Locator
HTTP   Hypertext Transfer Protocol
NLP   Natural Language Processing
IM   Interaction Manager
MC   Modality Component

References

Alias-I 2015, *LingPipe* [Online]. Available: http://alias-i.com/lingpipe/ [17 March 2015].
Api.ai 2015, *api.ai* [Online]. Available: http://api.ai/ [17 March 2015].
Barnett, J, Bodell, M, Dahl, DA, Kliche, I, Larson, J, Porter, B, Raggett, D, Raman, TV, Rodriguez,
     BH, Selvaraj, M, Tumuluri, R, Wahbe, A, Wiechno, P & Yudkowsky, M 2012, *Multimodal
     Architecture and Interfaces* [Online], World Wide Web Consortium. Available: http://www.
     w3.org/TR/mmi-arch/ [20 November 2012].
Barnett, J 2016, Introduction to the Multimodal Architecture, in *Multimodal Interaction with
     W3C Standards: Towards Natural User Interfaces to Everything*, ed, D Dahl, New York,
     Springer, to appear.
Claburn, T 2009, *Stephen Wolfram's Answer To Google* [Online]. Available: http://www.
     informationweek.com/news/internet/search/215801388?pgno=1 [9 November 2012].
Dahl, DA 2010, Extensible Multimodal Annotation (EMMA) for Intelligent Virtual Agents,
     10th Annual Conference on Intelligent Virtual Agents. Philadelphia, PA, USA.

Dahl, DA 2013a, Natural language processing: Past, present and future, in *Mobile Speech and Advanced Natural Language Solutions*, eds, A Neustein & J Markowitz, Springer.

Dahl, DA 2013b, 'The W3C multimodal architecture and interfaces standard' *Journal on Multimodal User Interfaces*, pp. 1–12.

Ferruci, DA, Brown, EW, Fan, J, Gondek, AK, Lally, A, Murdock, JW & Nyberg, E 2010, 'Building Watson: An overview of the DeepQA Project', *AI Magazine*, vol. 31, no. 3, pp. 59–79.

Fielding, RT & Taylor, RN 2000, Principled design of the modern Web architecture, *Proceedings of the 22nd International Conference on Software Engineering*. Limerick, Ireland, ACM, pp. 263–272.

Gildea, D & Jurafsky, D 2000, Automatic labeling of semantic roles, *Proceedings of the 38th Annual Conference of the Association for Computational Linguistics* (ACL-00), Hong Kong, ACL, pp. 512–520.

Gorin, AL, Riccardi, G & Wright, JH 1997, 'How may I help you', *Speech Communication*, vol. 23, pp. 113–127.

IBM 2015, *IBM Watson Developer Cloud* [Online]. Available: http://www.ibm.com/smarter planet/us/en/ibmwatson/developercloud/ [17 March 2015].

Johnston, M, Dahl, DA, Denny, T & Kharidi, N 2015, *EMMA: Extensible MultiModal Annotation markup language Version 2.0* [Online]. World Wide Web Consortium. Available: http:// www.w3.org/TR/emma20/ [16 December 2015].

Johnston, M, Baggia, P, Burnett, D, Carter, J, Dahl, DA, Mccobb, G & Raggett, D 2009a, *EMMA: Extensible MultiModal Annotation markup language* [Online], W3C. Available: http://www. w3.org/TR/emma/ [9 November 2012].

Johnston, M 2016, EMMA, in *Multimodal Interaction with W3C Standards: Towards Natural User Interfaces to Everything*, ed, D Dahl, New York, Springer, to appear. .

Johnston, M, Dahl, DA, Kliche, I, Baggia, P, Burnett, DC, Burkhardt, F & Ashimura, K 2009b, *Use Cases for Possible Future EMMA Features* [Online], World Wide Web Consortium. Available: http://www.w3.org/TR/emma-usecases/.

Jurafsky, D & Martin, J 2008, *Speech and language processing: An introduction to natural language processing*, Prentice-Hall, Upper Saddle River, NJ, USA.

Kelly, S 2015, *nlp_compromise* [Online]. Available: https://github.com/spencermountain/nlp_ compromise [18 March 2015].

Lee, K, Palsetia, D, Narayanan, R, Patwary, MMA, Agrawal, A & Choudhary, A 2011, Twitter Trending Topic Classification, *Proceedings of the 2011 IEEE 11th International Conference on Data Mining Workshops,* IEEE Computer Society.

LinguaSys 2015, *LinguaSys* [Online]. Available: https://www.linguasys.com/ [17 March 2015].

Loper, E & Bird, S 2002, 'Nltk: The Natural Language Toolkit', *ACL Workshop on Effective Tools and Methodologies for Teaching Natural Language Processing and Computational Linguistics*, 40th Annual Meeting of the Association for Computational Linguistics, July 11–12, 2002, Philadelphia, PA, USA.

Microsoft 2015, *Language Understanding Intelligent Service (LUIS)* [Online], Micosoft. Available: http://www.projectoxford.ai/luis [5 June 2015].

NLTK.org 2015, *Natural Language Toolkit* [Online]. Available: http://www.nltk.org/ [17 March 2015].

NLULite 2015, *NLULite* [Online]. Available: http://nlulite.com/ [13 March 2015].

Nuance Communications 2015, *Nina – The intelligent virtual assistant* [Online], Nuance Communications. Available: http://www.nuance.com/for-business/customer-service-solutions/ nina/index.htm [16 March 2015].

OpenEphyra 2015, *OpenEphyra* [Online], Carnegie-Mellon University. Available: https://mu.lti. cs.cmu.edu/trac/Ephyra/wiki/OpenEphyra [18 March 2015].

OpenNLP 2015, *Apache OpenNLP* [Online], The Apache Software Foundation. Available: http://opennlp.apache.org/ [17 March 2015].

Openstream I 2015, *Eva, Face of the Digital Workplace* [Online], Openstream. Available: http://www.openstream.com/eva.htm [16 March 2015].

Pandorabots 2015, *Pandorabots* [Online], Available: http://www.pandorabots.com/ [17 March 2015].

Schröder, M, Baggia, P, Burkhardt , F, Pelachaud, C, Peter, C & Zovato, E 2009, *Emotion Markup Language (EmotionML) 1.0* [Online], World Wide Web Consortium. Available: http://www. w3.org/TR/emotionml/.

Stanford Natural Language Processing Group 2014, *Stanford CoreNLP* [Online], Palo Alto, CA: Stanford University. Available: http://nlp.stanford.edu/software/corenlp.shtml.

Wallace, RS 2014, *AIML 2.0 Working Draft* [Online]. Available: http://alicebot.blogspot.com/ 2013/01/aiml-20-draft-specification-released.html [3 February 2015].

Wit.ai 2015, *wit.ai* [Online]. Available: https://wit.ai/ [17 March 2015].

Kathy Brown, Anmol Walia, Prashant Joshi, Abir Chakraborty, and Sanjay Agarwal

3 Omnichannel Natural Language

Bringing omnichannel and prediction to the multimodal user interface

Abstract: Consumers are increasingly expecting on-demand, natural, mobile solutions that enable them to describe their intent using natural language and to get information and services rendered transparently as they move across device and channel. In this work, we will discuss the process and benefits of combining a powerful, next generation user interface technology with Natural Language Processing, omnichannel prediction, big data and a real-time learning platform. This omnichannel Natural Language Understanding (Omni-NLU) solution delivers a powerful multimodal interface that integrates information across multiple modalities and channels – the web, agent conversations using chat or voice, social, SMS and mobile.

3.1 Introduction

The omnichannel NLU solution integrates the user's natural response extracted by Natural Language Processing (NLP) with omnichannel data to predict the optimum response for the user. The Omni-NLU interface ensures the user is able to move seamlessly across channels and modalities, maintaining context and always moving forward in getting answers and completing tasks without needing to back up or lose momentum.

In this chapter, we will describe:
– New designs for omnichannel, natural language modeling interfaces
– Technological advances in Natural Language Understanding
– The benefits of an omnichannel Natural Language Modeling for improving user experience and performance

3.2 Multimodal interfaces built with omnichannel Natural Language Understanding

We have developed a new omnichannel NLU technology that fuses multiple channels and modalities of data to build Natural Language, multimodal interfaces. In contrast to standard NL technologies (Cox & Shahshahani 2001; Sidorov, Velasquez, Stamatatos, Gelbukh & Chanona-Hernandez 2014; Jones & Galliers 1995), that process only the speaker's interaction from a single data channel, the new omnichannel NLU technology integrates the current interaction with past interactions across multiple channels to improve and inform the current NL interpretation. In this chapter, we will show how this technology works to infuse the current user interaction with contextual information to deliver an improved user experience and improved performance.

The omnichannel NLU technology enables the development of innovative multimodal interfaces that allow users to cross seamlessly across devices and modality. For instance, a user may start a conversation in a Virtual Assistant or even with a chat agent using text and then seamlessly transition to a voice conversation on their mobile device. The interface is driven by the user's preferred mode of interaction thus enabling the user to chat naturally using SMS as well as speak naturally using natural language. To illustrate the importance of a multimodal NLU solution, we illustrate in Fig. 3.1 multimodal consumer trends. In this graph, we see that 49 % of US consumers begin their customer service journeys on the PC and 32 % of consumers will go to the smartphone as their second device if they can't resolve what it is they need on the first device ([24]7 2015).

The Omni-NLU model is shown in Fig. 3.2. As shown, the Omni-NLU model is able to process inputs from numerous different modalities including text, voice and click data associated with online/web interactions. This technology allows the development of multimodal interfaces that allow the user to interact using a natural, conversational input irrespective of their modality. To the user, the interaction is natural and consistent with their modality of preference eliminating the need for the user to conform to a constrained input enforced by the technology. In this architecture, the Omni-NLU is smart enough to translate the input speech/text into a standard canonical form from which it performs the language understanding and intent prediction.

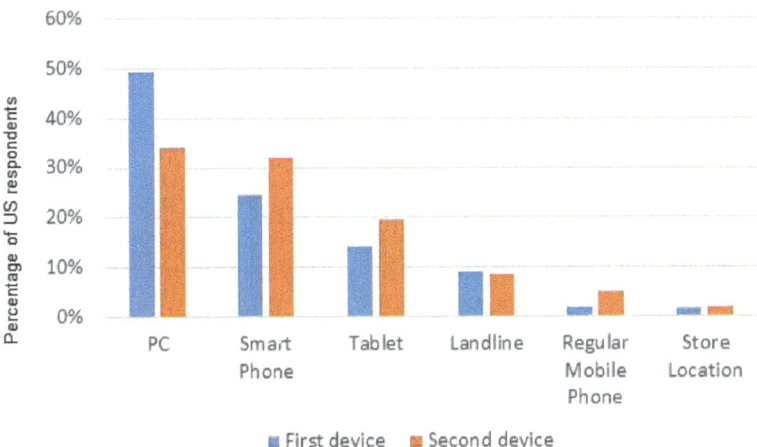

Fig. 3.1: Distribution of consumer multimodal usage statistics.

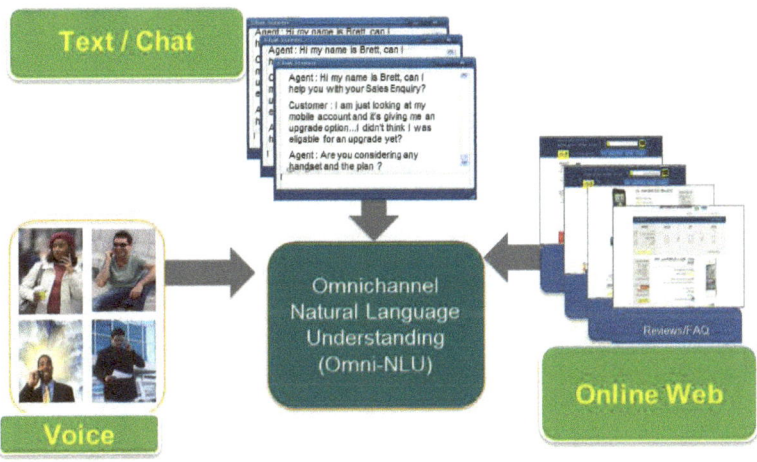

Fig. 3.2: Omnichannel Natural Language Understanding (Omni-NLU) architecture.

3.3 Customer care and natural language

Natural Language Understanding (NLU) is a field of Natural Language Processing that employs artificial intelligence, computer science, and linguistics to enable natural communication between people and computers. NLU technology analyzes the input text, extracts information, and attaches meaning to the words associated with the user interactions.

In the field of Customer Care, NLU is the underlying technology that allows users to interact with a system or device naturally and is at the core of a conversational user interface. Rather than constraining users to speak using a fixed syntax and grammar, NLU allows users to respond naturally to open-ended questions such as "How may I help you?" or "What's your question?"

In recent years, NLU has become an important technology at improving the user experience and containment rates of automated contact center solutions. NLU has allowed enterprises to move away from constrained interfaces such as the directed dialog IVR architecture illustrated in Fig. 3.3.

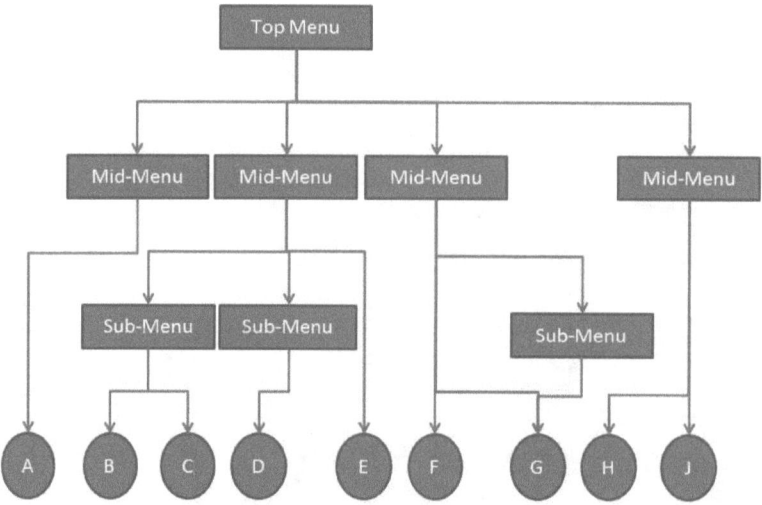

Fig. 3.3: Architecture of a standard directed dialog IVR.

As shown in Fig. 3.3, the standard directed dialog IVR consists of a hierarchy of static menus in which the user must interact with the system using a constrained syntax or grammar consisting of 2–4 words at each menu. The directed dialog IVR has proven to be extremely challenging for users to navigate and complete tasks successfully.

This interface has become even more challenged in recent years as users expect to do customer care on the go using multiple devices, channels, and modalities. As shown in Fig. 3.4, we have found that the omnichannel user will perform a greater percentage of simpler tasks online ([24]7 2015). In fact, we find that 64 % of US consumers begin their customer service journeys on a website and 32 % of consumers will go to the phone channel as their second channel if they can't resolve what it is they need in the first channel. As a result, the IVR must now be

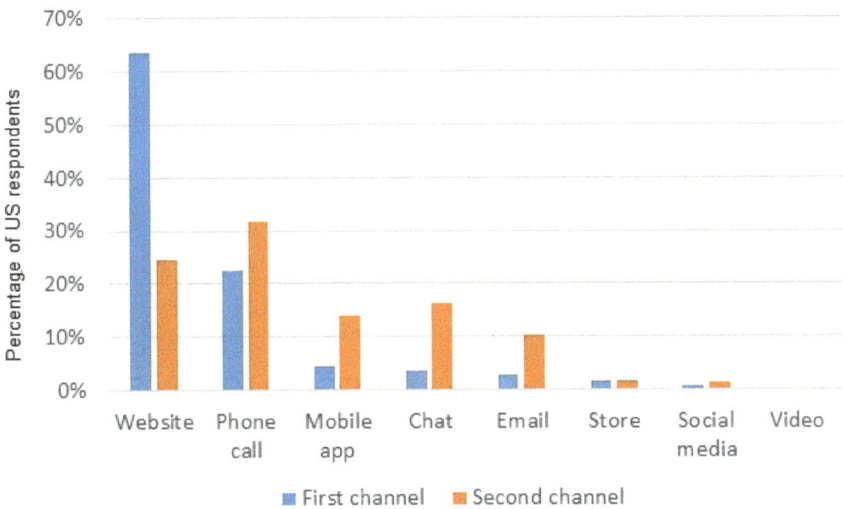

Fig. 3.4: Distribution of omnichannel usage statistics.

able to handle the more complex requests and sentences that consumers use to re-solve difficult tasks or to resolve an escalated issue. In fact, the standard IVR menu structure has quickly become obsolete and stale and is not meeting the needs of the new omnichannel user.

To highlight this disconnect further, our recent analyses of directed dialog systems indicate there are increasingly different usage statistics. Specifically, in our recent analyses, we find that users are increasingly asking for the "more options" selection at the main menu as they are not finding what they need in the constrained main menu of the standard IVR. Forcing users to navigate a "more options" menu tree to find their selection has proven to be very frustrating to users resulting in high abandonment and agent transfer rates.

In Fig. 3.5, we highlight the differences between a directed dialog and NLU architecture. In an NLU system, the user can speak naturally and state the reason for their call or interaction. The NLU system enables users to navigate IVR trees more quickly, even having the ability to reach terminal nodes through a one-step dialog.

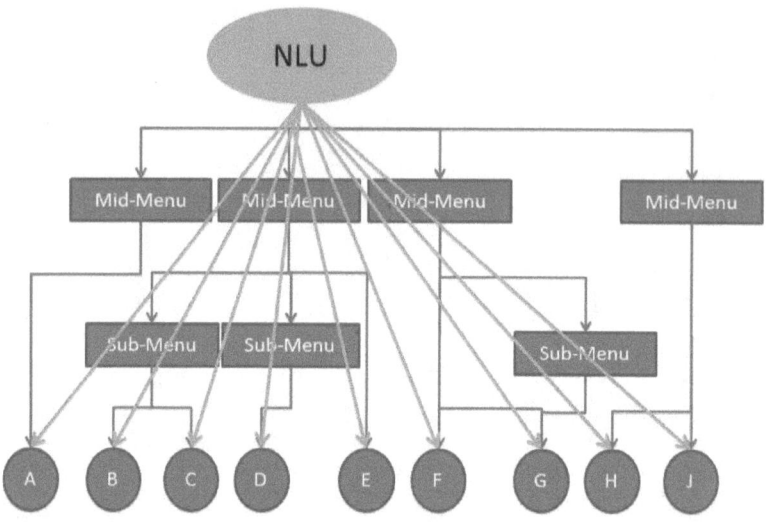

Fig. 3.5: Architecture of a standard NLU IVR.

3.4 Limitations of standard NLU solutions

While the standard NLU architecture has improved Customer Care KPIs, associated with CSAT and containment rates, this standard NLU approach has also encountered limitations that have prevented the technology from improving further. In this chapter, we present the new omnichannel NLU architecture to address and overcome these limitations.

Language is a complex formulation of sounds, syntactic patterns, and semantic inferences. Within the research community, NLU is classified into three major phases of analysis of increasing complexity comprising syntactic, semantics and pragmatics processing.

Syntactic processing deals with the rules for how to combine words into grammatically correct phrases and sentences. Semantics considers the meaning of words, phrases, and sentences and the associations by which meaning and concepts are attached and communicated through natural language. Pragmatics deals with the ways in which language is interpreted through the larger domain of discourse and past experiences.

Within the research community, it has been established that in order to develop an NLU system that approaches human comprehension levels, we must incorporate pragmatics and discourse into the interpretation (Allen 1987). Such advanced processing is necessary because it has been shown that conversational language, the "art of discourse", is non-specific. Speakers frequently provide incomplete descriptions of the information they want to convey. The words and phrases they use can be vague and frequently will leave out important information or details that the system is "expected" to know. Additionally, new words, expressions and meanings evolve. Languages evolve as experiences change. In the area of human comprehension, we do not encounter these types of limitations as language and the words that comprise sentences and phrases elicits a stream of semantically related concepts, past experiences and related events that enable us to complete complex NLU tasks – such as word sense disambiguation, textual entailment, and semantic role labeling – in a quick and effortless way.

Existing NLU technologies that have been deployed in contact center solutions employ the most basic syntactic paradigms of interpretation. In these systems, syntactic representations of text are based largely on keyword spotting, page rank algorithms (Page et al. 1999), or statistical word language models (Joachims 2002) to perform interpretation and extract intent. Such algorithms are limited in resolving ambiguity because they can process only the information that they can "see" or "hear." They do not process ideas outside of the information derived from keywords or their synonyms within the input. They cannot create a new idea based on the information posited from previous inputs.

Consequently, the standard unichannel NLU system presented in Fig. 3.5 and deployed in most contact center solutions encounters a barrier beyond which they cannot improve performance because they cannot overcome the inherent ambiguities of human language and discourse. To demonstrate the limitations of the standard NLU system, we depict a typical usage scenario in Fig. 3.6. In this use case, the user has responded to the open-ended question "How may I help you?" with a vague response such as "I'm having a billing problem." In this case, the NLU system can understand the words in the sentence, but clearly cannot deduce the specific problem due to the ambiguity. As a result, it drops the user down into a directed dialog node forcing the user to describe their problem using a fixed syntax grammar, which can be a formidable task. Due to high frustration levels, users will drop out or transfer immediately to an agent.

To further illustrate this limitation, we provide in Fig. 3.7 a chart that shows a typical intent distribution for the standard NLU system. In this distribution, we see that a large percent of the intents are non-meaningful, i.e. vague intents, out of domain intents and requests for an agent. In a standard NLU system, these intents require further disambiguation usually ascertained through the standard directed dialog menus.

In order to overcome these limitations and advance the state-of-the-art in NLU, we present the new omnichannel NLU technology that improves intent understanding by leveraging the larger context of the user. General knowledge about the domain is employed as well as specific knowledge about the user or sit-

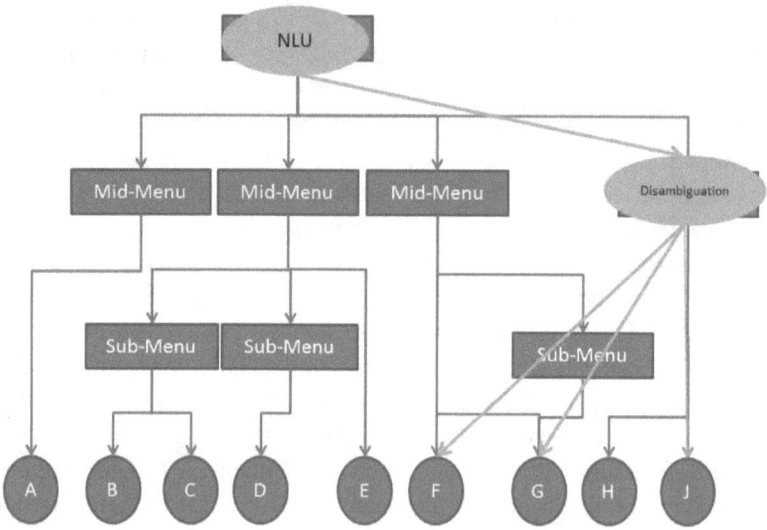

Fig. 3.6: Limitations of the standard, unichannel NLU solution.

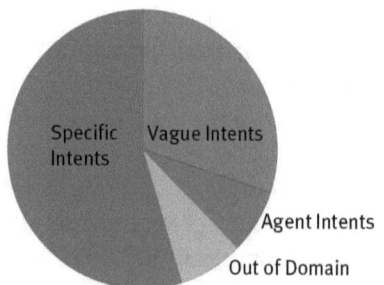

Fig. 3.7: Intent distributions for a standard NLU system.

uation. This knowledge is critical to understanding the intentions of the speaker and to enable the specification of the background assumptions presumed by the speaker. As an example of this ability to capture ambiguity, consider the sentence "I'm having a problem with my bill." The sentence is ambiguous and cannot be disambiguated without considering the larger discourse context, not available at the level of a syntactic or semantic approach. We need a mechanism by which we can also incorporate background knowledge and reasoning to fully understand and resolve the intent.

The Omni-NLU technology improves intent resolution by moving away from a purely syntactic approach to one that incorporates semantics and pragmatics by integrating NLU technology with context-dependent information. In the Omni-NLU system, we predict the intent of the user's current interaction given their

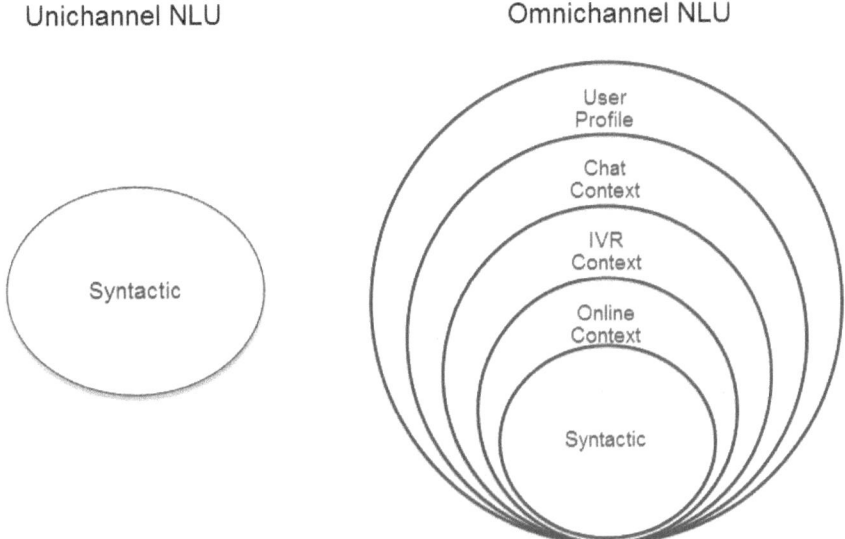

Fig. 3.8: Unichannel NLU vs. omnichannel NLU models.

previous interactions across multiple channels and modalities. This technology is critical in enabling the development of successful multimodal interfaces for the new "mobile" user who moves across devices and modalities to achieve goals and execute journeys.

In Fig. 3.8, we illustrate the key differences between the standard unichannel and new omnichannel NLU models. In the new omnichannel NLU technology we include a set of hierarchical context-dependent predictive models integrated with the standard syntactic NLU model. The additional layers model the context associated with multiple channels and modalities. In this new NLU technology, we incorporate the background of the user – their objectives, goals, expectations – into the interpretation of their language to guide, inform and improve our current understanding. Using this new omnichannel NLU technology, we move from the domain of syntactic and "semantic interpretation" to "contextual interpretation."

3.5 Omnichannel NL architecture

The architecture for the Omni-NLU technology is illustrated in Fig. 3.9. The architecture consists of several layers of processing to interpret and extract the intent of the user.

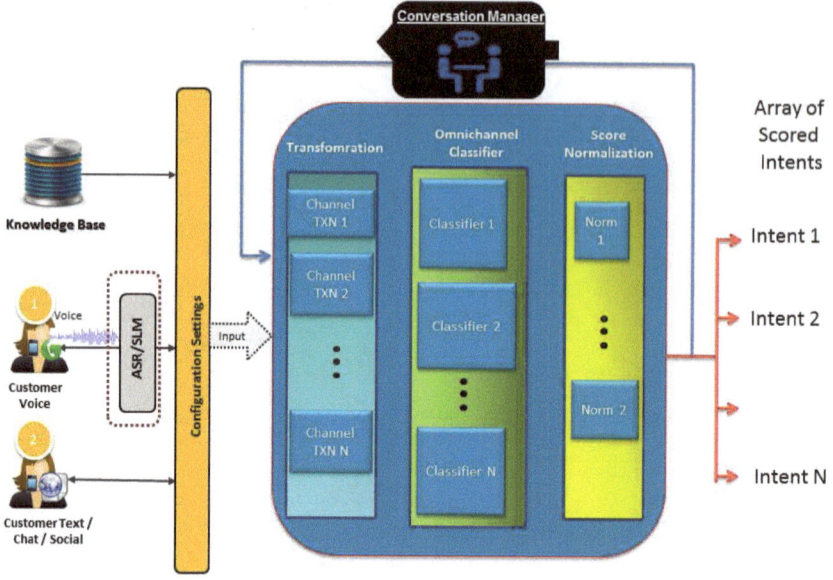

Fig. 3.9: Omnichannel natural language architecture.

Key features of the architecture include the following:

– Seamless NLU architecture that can process both structured and unstructured data across multiple channels including text, chat and voice.

– Development of multimodal interfaces driven by one omnichannel NLU model that can effectively process unstructured data from multiple modalities such as voice, chat, SMS and social.

– A bank of predictive omnichannel classifiers that can process and extract contextual information, meaning and pragmatics from a series of channels including:
  – Chat
  – Online/Web
  – IVR

– Omnichannel normalization algorithms for transforming and tokenizing the input from different channels such as voice and chat into a standard logical form.

– Hierarchical time-depending processing architecture that can enable the development of multiple levels of conversational NLU.

– One single architecture for building NLU solutions across multiple modalities including IVR, Virtual Assistants, Mobile, Chat Assist, and Voice Agents.

– Unified classifier architecture for combining structured and unstructured data.

Each of the components of the new omnichannel Natural Language Understanding technology are described in the sections below.

3.5.1 Omni-NLU training algorithm

The training process for the Omni-NLU model is illustrated in Fig. 3.10. As shown, the Omni-NLU model employs machine learning algorithms to fuse multiple channels of data into a single unified model. This is an innovative approach that offers the following advantages:

- **Leverage voice and chat data at scale** to evolve natural language models more quickly, resulting in better performance (greater recognition accuracy and intent understanding) and faster time-to-benefit
- Make natural language a fundamental component of omnichannel customer engagement
- Apply **one language model** across different channels such as chat interactions, mobile apps and virtual assistants
- Create **consistent and intuitive** natural language experiences
- **Lower TCO** and reduce deployment risk for natural language

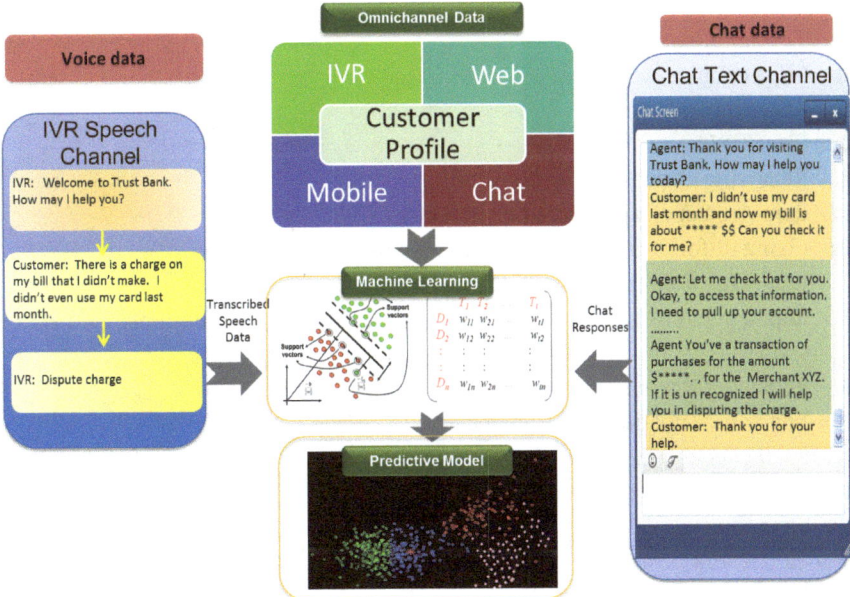

Fig. 3.10: Omnichannel Natural Language Understanding training architecture.

3.5.2 Statistical-language model

Objective: The Statistical-Language Model (SLM) is a statistical grammar that consists of a trigram language model. The SLM is used by an Automatic Speech Recognition (ASR) engine to convert the audio signal into a string of words. It is built to recognize open-ended responses provided to natural language prompting. The SLM is trained using offline tools from large corpuses of data that include transcribed speech and chat data.

3.5.3 Input transformation

Objective: The objective of the transformation (TN) module is to preprocess the data prior to performing the classification. In the transformation phase, the text is transformed into a standard form including modules such as spell checkers and stemming.

Different types of text will need different types of text normalization; and as a result this block allows flexibility in the configuration and selection of components and their order. For example, chat data will have misspellings, special characters; however, voice data decoded by the ASR will not and so will not need all the transformation layers.

The modules include:
1. Class email
2. Class time
3. Class URL
4. Removing extraneous symbols
5. Removing special formatting characters
6. Fixing ill-formed dates and replacing dates
7. Replacing dollars and symbols
8. Replacing multiple punctuations and removing punctuations, and extra spaces
9. Replacing the remaining numbers
10. Replacing substitutions from a list
11. Spell checking
12. Replacing word classes
13. Stemming
14. Stop words
15. Regular expressions

3.5.4 Predictive omnichannel classifier

Objective: The predictive omnichannel classifier is a multistage, hierarchical algorithm that processes the input to extract the intent or meaning. The hierarchical algorithm employs a bank of classifiers intended to fuse the different modalities and channels of input. All training of models occurs using offline processes. This block outputs a set of intents and scores in an ordered list.

3.5.5 Score normalization

Objective: The Score Normalization module converts the scores from the omnichannel classifier into a different scale from which intents can be relatively compared and rejected using thresholds. This component is important because the applications that consume the classifier output will need to use thresholds to control the conversation and to determine when error recovery dialog is required.

3.5.6 Conversation manager

Objective: The Conversation Manager performs dialog management. It will maintain the state of the conversation and determine what stage of the classifier hierarchy should be applied to any node in the conversation. The data that will be input to the CM is the Context Block.

The Context Block (CB) is an archive in which previous intrachannel interaction node information is stored. The following information is stored in the CB:
– Input text (chat or decoded voice)
– Output from classifier including the intent matrix and associated scores

This NL approach is quite different from existing natural language technologies. As illustrated, the multistage, hierarchical classifier integrates additional omnichannel streams of data to derive the intent of the user.

This approach yields a stronger natural language technology that fuses the standard input text or speech of the user with additional context-dependent blocks obtained across numerous different interactions and channels to predict the intent of the user. In the standard NLP approaches, the text of the user is the only input that is used to extract the intent of the user.

3.6 Experimental results

We present the results of applying the Omni-NLU technology in a Customer Care solution that has been deployed in the financial services sector. In this application, users are offered several different types of self-service options or can get directed to an agent for further interaction if the system cannot resolve the intent.

3.6.1 Current analysis segment

For the present analysis, we have considered calls that happened between January 1, 2015 and February 15, 2015 (inclusive of both the days). In this period, there were a total of ~ 6.5 M calls. Shorter duration calls (less than 30 seconds) or calls that resulted in hang-up (and not transferred to an agent) are further removed from consideration. This resulted in 5.98 M calls.

In this study, we evaluated the performance of the system comparing the omnichannel NLU with the standard unichannel NLU performance. The unichannel NLU solution does not use context-dependent data in the training. In the case of the omnichannel NLU model, context dependent information was incorporated into the model associated with previous inter- and intrachannel interactions.

In this study, we have shown that the Omni-NLU provides a 38 % improvement in classification performance over the unichannel model on calls in which the omnichannel classification is enabled. The results are most compelling for resolving vague, ambiguous and agent requests. For some intents, such as charge query and dispute intents, the user frequently will respond to the open-ended NLU question by asking for an agent giving no opportunity for automation. In fact, these users get stuck in a repeated cycle of interactions as they get information about the charge, file a dispute and then check status for resolution. As an example, in Figs. 3.11 and 3.12 we provide the intent distributions for repeated interactions wherein the users ask for an agent in the current interaction given the previous interaction was a Charge Query intent.

Particularly notable is that when the current interaction occurs within six hours of the previous interaction, the consumer is attempting to notify the company that they are having a problem with a charge and want to file a dispute or file a fraud report (Dispute Notify/Fraud Report). However, when the current interaction occurs longer than a day after the previous interaction, the usage statistics change wherein the intent of getting the status of a dispute has doubled from that in the distribution for six hours. This changing trend reflects the journey as consumers progress through the task of resolving problematic charges.

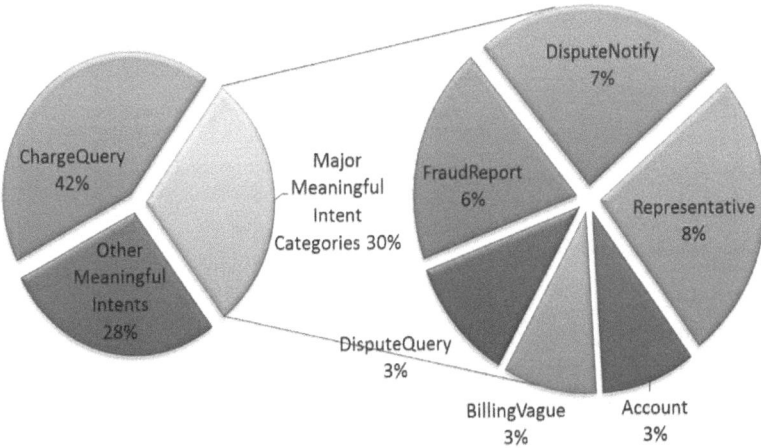

Fig. 3.11: Distribution of intents for interactions that occur within six hours of the previous interaction where the previous interaction was a Charge Query intent.

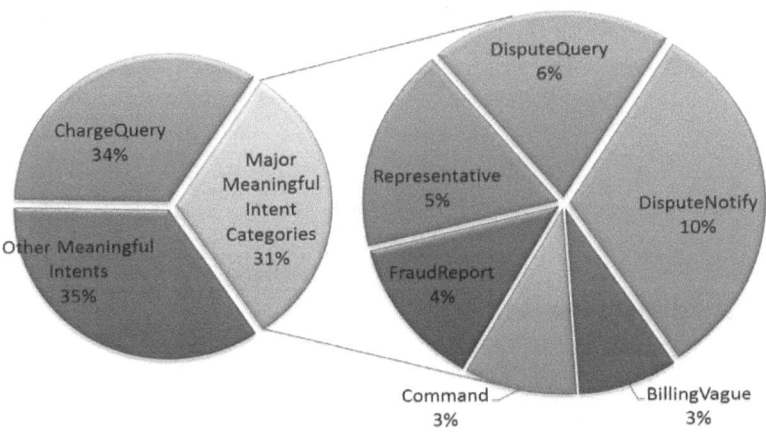

Fig. 3.12: Distribution of intents for interactions that occur after one day of the previous interaction where the previous interaction was a Charge Query intent.

Figure 3.13 is an illustration of how the Omni-NLU technology transforms the user's input, fusing it with contextual information, to yield a more accurate understanding of the user's intent.

In this example, the user's input is combined with data from other channels to extract the current real-time intent and provide the user with a recommendation.

Through the transformation, we are able to immediately provide a recommendation to the user on how to quickly get a solution and resolve their issue. In

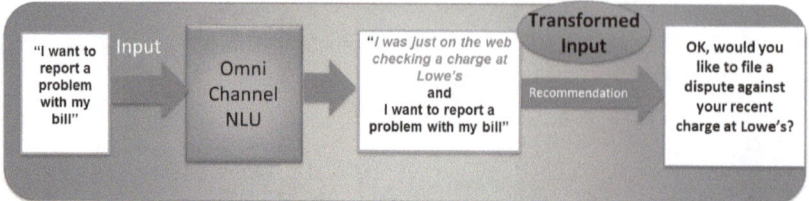

Fig. 3.13: Omnichannel transformation.

contrast, a standard NL solution would have to take the user through additional disambiguation menus to attempt to determine the source of the problem the consumer is attempting to get assistance with.

If we look at the classifier view illustrated in Fig. 3.14, we see that the omnichannel classifier performs a multilayer transformation. The first transformation, in the upper left, is the transformation performed at the syntactic layer. That transformation then undergoes a subsequent transformation T' by fusing it with omnichannel data.

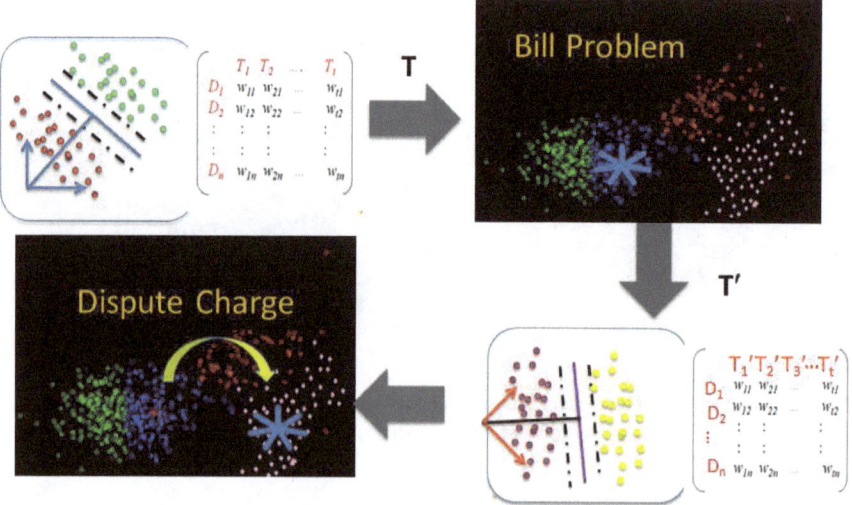

Fig. 3.14: Omnichannel classifier transformation.

3.7 Summary

To conclude, we have presented a novel technique for building multimodal user interfaces that integrates the user input with additional streams of omnichannel data to improve the accuracy of a natural language interface.

This approach yields a stronger natural language technology that fuses the standard input text or speech of the user with additional behavioral characteristics obtained across numerous different interactions and channels to predict the intent of the user.

Abbreviations

Omni-NLU	Omnichannel Natural Language Understanding
NLU	Natural Language Understanding
SLM	The Statistical Language Model
IVR	Interactive Voice Response
ASR	Automatic Speech Recognition
TN	Transformation
CB	Context Block
TCO	Total Cost of Ownership

References

Allen, J 1987, *Natural Language Understanding*, Benjamin/Cummings, Redwood City, CA.

Cox, S & Shahshahani, B 2001, 'A comparison of some different techniques for vector based call-routing', *Eurospeech 2001*, Scandinavia.

Joachims, T 2002, *Learning to classify text using support vector machines: Methods, theory and algorithms*, Kluwer Academic, Norwell, MA.

Jones, K & Galliers, J 1995, *Evaluating Natural Language Processing Systems: An Analysis and Review*, Springer, Heidelberg, Germany.

Page, L, Brin, S, Motwani, R & Winograd, T 1999, 'The pagerank citation ranking: bringing order to the web,' Stanford Univ., Stanford, CA Tech. Rep.

Sidorov, G, Velasquez, F, Stamatatos, E, Gelbukh, A & Chanona-Hernandex, L 2014, 'Syntatic N-grams as machine learning features for natural language processing', *Expert Systems with Application*, vol. 41, issue 3, pp. 853–860.

[24]7 2015, 'Customer Engagement Index', [24]7 Inc., Campbell, CA.

Michael Lawo, Robert K. Logan, and Edna Pasher

4 Wearable computing

A media ecology approach and the context challenge

Abstract: We describe the operations, features, and implications of wearable computing (a computing system worn by the user like clothing). The interactional considerations between users and wearable computing are examined following the ideas of J. C. R. Licklider, Andy Clark, Marshall McLuhan, Neil Postman and the concept of Media Ecology. We argue that understanding the environment created by wearable computing as well as its psychological and social impact is crucial to approach this technology. As McLuhan pointed out, one cannot understand a figure unless one understands the environment in which it operates. To this end, we introduce the notion of Wearable Ecology as a way of understanding these impacts. Furthermore, we review some past achievements and the state of the art with outcomes of latest research on context frameworks for wearable computing. Finally, we discuss considerations for future directions for the development of this technology.

4.1 Introduction to Wearable Ecology

A wearable computer is a computation system consisting of different components that is worn by the user, either directly as an integrated device such as a wristwatch (smartwatch), or integrated into clothing. In order to understand the impact of wearable technologies we will first introduce the notion of wearable ecology as an offshoot of media ecology that emerged as a result of the work of Marshall McLuhan and Neil Postman. The idea behind media ecology is that independent of its content, a medium has certain effects by virtue of the way in which the user interacts with that medium. This view is contained in McLuhan's signature one-liner "the medium is the message". In other words, in addition to the message conveyed by content, there is another message conveyed by the medium itself. That is, a given medium with all its affordances and limitations produces effects on users, independent of content. We apply this notion of media ecology to wearable technology and describe it as Wearable Ecology. Figure 4.1 depicts some examples of Wearable Ecology.

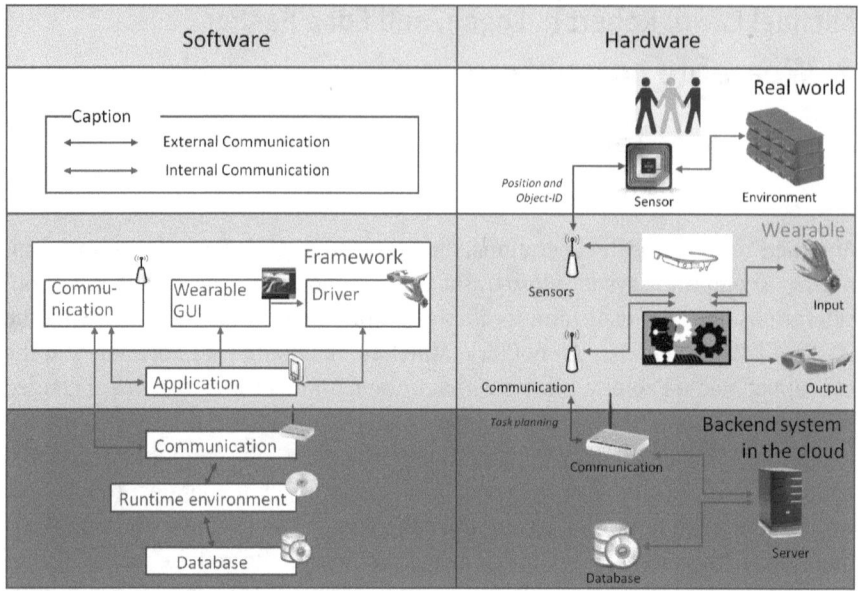

Fig. 4.1: Wearable Ecology – examples.

Our notion of Wearable Ecology draws significant inspiration from the work of J. C. R. Licklider, Andy Clark, Marshall McLuhan, and Neil Postman. Underlying this notion is the claim that there is an intimate relation between a user and a given technology, which creates an interesting paradigm we call Wearable Ecology. The following quotes contribute to the basic framework for our discussion on Wearable Ecology:

> All media are extensions of some human faculty – psychic or physical.
> (McLuhan & Fiore 1967, p. 26)

> Technologies are merely extensions of ourselves.
> (McLuhan 1967, p. 261)

> To behold, use or perceive any extension of ourselves in technological forms it is necessarily to embrace it. By continuously embracing technologies, we relate ourselves to them as servo-mechanisms.
> (McLuhan 1964, p. 46)

According to McLuhan (1964), our technologies are extensions of our bodies and our communication media are extensions of our psyches or minds. Andy Clark (2003), who developed the Extended Mind Hypothesis, states that our mind is not simply constricted inside the skull but rather, that our tools are actually part of our minds. Bringing McLuhan and Clark together, we arrive to the notion that wearable technologies are then not only a continuation of our mind, à la McLuhan, but also that these tools effectively expand our cognition. According to Clark, we are "natural born cyborgs" (2003) that use tools to extend our minds. Along this line, the designers of wearable technologies can be seen as using their minds to create further extensions of human minds. Just as McLuhan had once suggested when he wrote, "Man becomes ... the sex organs of the machine world ... enabling it to fecundate and to evolve ever new forms" (McLuhan 1964, p. 56). Wearable technologies are second order extensions of our brains/minds in which the software of the brain is used to create a combination of hardware and software to further extend the perceptual hardware of the body and thereby extend the software of the mind. In other words, wearable technology is an extension of an extension – or a second order extension. The question that surfaces then is: Does this actually benefit the user and if so, how?

Every technology brings both service and disservice. For instance, smartphones extend the reach of our psyches outside the range of our immediate physical surround in the same way that the telegraph and telephone (and later also email and video conferencing systems) extended the reach of communication and allowed collaboration and coordination among individuals that were not able to communicate face to face. The speedup of electric communication has created a new kind of human with enhanced communication capabilities but also with new challenges, such as information overload and the reduction of attention span (McLuhan 1964 and Postman 1990). What will then be the effect of the speedup created by wearable technology?

Needless to say: The resulting greater efficiency is a positive outcome. However, the quality of the decision-making process that takes place instantaneously may encourage us to make decisions that are not appropriately thought out. That is, while greater efficiency may improve our decision-making process, the rush to judgement encouraged by the speedup of wearables may degrade our thought processes. We already suffer from information overload (Postman 1990) so that attention economy now becomes another issue we have to pay attention to, even though the speedup of technology may not necessarily increase the quality of our decision-making. Just as the heart can go into the arrhythmia, so can the brain go into a state of confusion if over stimulated – a state of mania. Perhaps, what is needed is a wearable technology that reminds us when we are overloading our brains and when we should take "the damn thing off" for a certain period of time. For instance, the device could automatically turn off when it senses the user is experiencing information overload by reading signals of the user's brain activities. Or perhaps, the wearable could warn us if we are reacting too quickly to a new input and thus prompt us to think again before we react to incoming information. A message could appear once a response is made too quickly after receiving some new input. The message could read as follows: "You responded too quickly to that input and therefore after 30 minutes pass you will be asked to review your response to see if you still want to send it". An option to override this prompt should also be integrated in case of an emergency.

Clark (2003) argues that we are natural born cyborgs in that from the first emergence of Homo sapiens, we have always used technology to extend our capacity to achieve our goals and as a result have enriched our existence and extended our life span. However, the question of where to set the limits to this expansion remains unanswered. In order to address this issue, we need to look at both the services of wearables as well as their potential disservices so we can maximize the former and minimize the negatives. This view of humans using technologies to extend skills and resourcefulness deeply relates to the notion behind human-computer symbiosis.

4.2 Human-computer symbiosis

The idea of a human-computer symbiosis that enhances the human capacities dates back to "computing's Johnny Appleseed" Licklider's paper in 1960. The author understands human-computer symbiosis as a cooperative interaction between the user and the digital device:

> (...) involving a very close coupling between the human and the electronic members of the partnership aiming at computers facilitating formulative thinking as next step beyond solving formulated problems enabling men and computers to cooperate in making decisions and controlling complex situations without inflexible dependence on predetermined programs. In the anticipated symbiotic partnership, men set the goals, formulate the hypotheses, determine the criteria, and perform the evaluations. In the background the computing machines do the routinizable work that must be done to prepare the way for insights and decisions in technical and scientific thinking. Preliminary analyses indicate that the symbiotic partnership will perform intellectual operations much more effectively than man alone can perform them.
> (Licklider 1960, p. 4)

Licklider distinguishes between human extended machines, where the human operator does things not possible to automate, and the human-computer symbiosis where "the computing machine [is] effectively [brought] into the formulative parts of technical problem" (Licklider 1960, p. 5). Licklider's paper is visionary and worth reading still today. His ideas about the mismatch between humans and computers, memory requirements and its organization, as well as the language problem have to be understood from the computing perspective of the late 1950s. For Licklider, we should instruct the computer by goals using mathematical programming to achieve them, involving real-time concatenation of preprogrammed segments controlled by the human operator.

Most interesting is what this author requires for input and output equipment: First, there is the interactive desk-surface display and control where the user is required to provide written or sketched information to a computer as would be done for another human (today implemented as Terminal/PC/Notebook/Smartphone). The second equipment is a computer-posted wall display (today implemented as Projector/Screen) to share information with individuals using the interactive desk-surface display. As a third component Licklider sees speech interaction as an automatic speech production and recognition system as the most natural mode of communication with the computer. At the time of writing his paper he was aware that speech synthesis (speech production as output) was available. There were already the first successful experiments with speech recognition (input) and he expected a sufficient capability to interact (input and output) with the computer by speech to be available by the mid-1960s. In the next section, we turn to the challenges and achievements of speech recognition and other interactional features of wearable technology.

4.3 Interactional considerations behind wearable technology

Speech is one of the main desired technologies for interaction in wearable technology, and it poses a great challenge for developers. Whereas the first two categories of equipment for input and output respective interaction have been achieved in our everyday work in the human-computer symbiosis, the third one (speech recognition) has not, as the challenges were by far underestimated. Below are the main reasons behind the lacking capabilities.

1. The communication between humans is redundant and benefits from the great variety of possibilities to express ourselves with words, gestures, mimics, intonation, etc. Forsberg (2003) provides revealing examples of issues for misunderstandings in speech-based human-computer communication using automatic speech recognition software.

2. Humans experience frustration and change their way of speaking if the computer does not understand what is said a couple of times in a row. Forsberg (2003) even doubts that perfect automatic speech recognition is ever possible and emphasizes especially the matter that humans speak differently to computers than to humans.

3. Users have to adapt to the computer to increase the quality of speech recognition for efficient user interfaces. This limits human capabilities and hinders creativity.

The display to view and to share information that Licklider described are today available even on the move in the form of smartphones connected with the Internet. However, providing a system with our ideas and intentions, for example during assembly or maintenance tasks, is a much more involved process as the problems encountered in speech recognition described above show.

The specific contexts of the situation in which humans find themselves, as well as procedural and tacit knowledge are the minimum requirements. These requirements, nevertheless, can become complex. Such is the case with the requirements and procedures described in a maintenance manual of an aircraft required to follow when performing a maintenance task.

Here, Cyber Physical Systems (CPS) like sensors, actors or displays can provide information to the networked system automatically, dynamically and implicitly (i.e. applications running on a computer network). This is e.g. beneficial in human-machine or robot collaboration at an assembly line or in an emergency response setting when tracking a firefighter during an operation by the command post. Also in healthcare settings, CPS yield medical data on the move to a backend system in a hospital or ambient assisted living setting. In all these examples, the worn or in clothing integrated CPS with the ability of connectivity to the Internet results in an added value compared to any mobile device. This is mainly because the CPS provides information in an ambient way thus unburdening the user of explicit (textual) input.

In fact, for Licklider although the computer (network) has the ability to store and perfectly retrieve information and to perform complex calculations, it requires human guidance in order to be creative and productive. CPS can then provide and utilize information in an automatic manner as described above by using context information and replacing explicit user input by automatically generated input. This might lead to the assumption that the user is no longer a guiding and creative force but solely productive like a robot controlled by a program. One could mean that in a situation in which the CPS knows everything in advance, a complete system could be set up and the human becomes a part of it, just as a robot or device for which creativity is required the least.

In his paper on proactive computing, David Tennenhouse touches upon some interactional considerations between users and computers. The author states that the number of networked interactive computers will surpass the number of people on the planet and asks: "How are things different when the interface is being used to supervise thousands of computers and/or millions of knowbots?"

(Tennenhouse 2000, p. 46). Apart from mainframes and personal computers, Tennenhouse also counts as networked interactive computers devices such as microcontrollers, embedded microprocessors, digital signal processors and computational microprocessors, which we all summarize today mostly as CPS within the Internet of Things (IoT). In CPS, computation is integrated into physical processes where physical processes affect computations and vice versa. However, to fully realize the potential of CPS, "the core abstractions of computing need to be rethought (Lee 2008, p. 1)."

Although Tennenhouse still sees that we have a long way to go in this regard, he claims that:

> The computer science research community now enjoys a rare and exciting opportunity to redefine its agenda ... In lifting our sights toward a world in which networked computers outnumber human beings by a hundred or thousand to one, we should consider what these "excess" computers will be doing and craft a research agenda that can lead to increased human productivity and quality of life
> (Tennenhouse 2000, p. 43)

Incremental improvements will continue to help. However, effective orchestration of software and physical processes requires semantic models that reflect properties of interest in both.

Tennenhouse requested three foci of research for proactive computing: (1) *getting physical* when using sensors and actuators, (2) *getting real* to routinely and directly respond without human interaction and (3) *getting out* where humans find their way *from within to above the loop* with the computer. What Tennenhouse realized was the following: Beyond the mainframe there were already many other computing devices for which we as users could benefit by a merge. This merge would allow synergies from the information gathered by sensors and actuators that respond to requests by starting an autonomous (automatic) dialog of sensors and actuators: this he called "proactive computing".

We now have software-friendly approaches with sufficient system flexibility and performance due to context-dependent control systems. From predicting the near-term performance, they create the context information under a range of possible inputs just-in-time using background processes for control. Control systems ensure operation within gross bounds known to be safe, leaving the detailed operation of the system to heuristic and/or statistical algorithms. Tolerating statistical variations in component availability and connectivity leads to new ways of thinking about fault tolerance in distributed control systems as an essential requirement.

Tennenhouse encouraged thinking about the interface to proactive computing and raised questions such as: "How should humans interface with systems whose response times are faster than their own? How are things different when the interface is being used to supervise thousands of computers and/or millions of knowbots?" (Tennenhouse 2000, p. 48) Here, replacing explicit human interaction by human controlled output of CPS (sensors) can be the key.

Also Humanistic Intelligence discusses Interactional aspects. In his significant paper (guest editor's introduction) on wearable computing, its pioneer Steve Mann explained how to understand this new interface as Humanistic Intelligence (HI) (Mann 2001). He claims to overcome the separation of human and computer as two distinct entities. He regards the computer as a second brain and its sensory modalities as additional senses, which synthetic synesthesia merges with the wearer's senses:

> When a wearable computer functions in a successful embodiment of HI, the computer uses the human's mind and body as one of its peripherals, just as the human uses the computer as a peripheral. This reciprocal relationship is at the heart of HI.
> (Mann 2001, p. 10)

The intent is to achieve synergy through a user interface to the signal-processing hardware that is in close physical proximity to the user and continuously accessible.

It is a matter of fact; the wearable makes us aware of peripheral information. Nevertheless, the relevance of this acquired information needs filtering. At the beginning, what we obtain is just data but what we need is information, knowledge or even wisdom for data to be actually helpful in a specific situation. Here we define information as contextualized data, knowledge as the ability to use information to achieve our goals, and wisdom as the ability to choose goals consistent with our values. We have furthermore to determine when it is wise to use wearables at all and when it is not. Wearables provide only information; and therefore, we should not overestimate this technology by allowing it to interfere with our judgement. This is of course a matter of context: The ability to use logarithmic tables properly is knowledge; but logarithmic tables are only information. The dream is "wearing a company" by having the wearable access a knowledge network of one's company or organization, however, what we obtain is information. In short, we look for the knowledge network but so far, the wearable technology can only offer an information network.

Another interactional issue regarding wearable technology is the amount of user participation it requires. McLuhan classified media as either hot or cool. The former refers to a high-definition communication that demands little involvement from the audience, whereas cool media describes media that demand active involvement from the audience. Following this rationale, we characterize wearable technology as a cool medium. It demands a lot of participation and although it is a smart technology, we must ask whether it is also a wise technology. Wearable technology only becomes a wise technology if it is used appropriately, which will require user training. Therefore, we have to provide training for wearable technology to wisely use: to know how to use it, when to use it, where to use it and when to turn it off or take it off.

When reviewing the evolution of technology from Mainframe to Desktop, through Notebooks to Tablets, to Smartphones and Wearables, we must consider the environment or ground created by the use of these tools. We use wearables in an ambient way to enhance the users' navigation through their interactions with the real world. Thus, the objective of any wearable computing application is to provide relevant and timely information as needed in an ambient way while at the same time the user is interacting with the real world and the wearable system. In mobile applications as in desktop applications, the user interacts either with the real world or with the system but never with both at the same time. Figure 4.2 below shows interactional characteristics of mobile versus wearable devices.

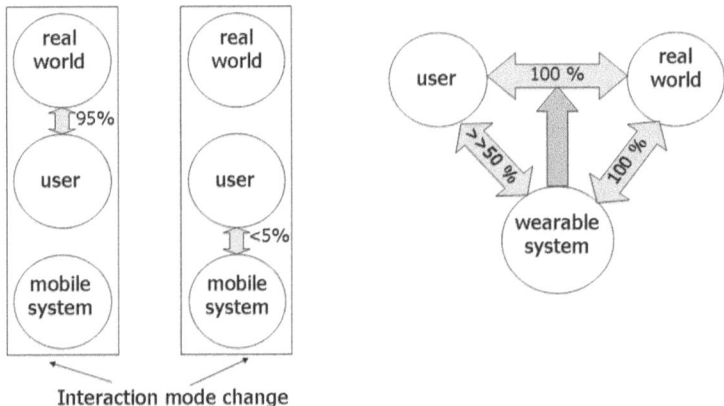

Fig. 4.2: Mobile versus wearable interaction.

One example of wearable technology use is during a maintenance task where the maintenance manual provides information during the operation on a head mounted display. The separation between gaining the relevant information and accessing it vanishes. There is no need to read a (paper-)manual first, nor is a training session to learn the task before required, as the information of what to do is available on the spot. The following Fig. 4.3 illustrates access to information during a maintenance task.

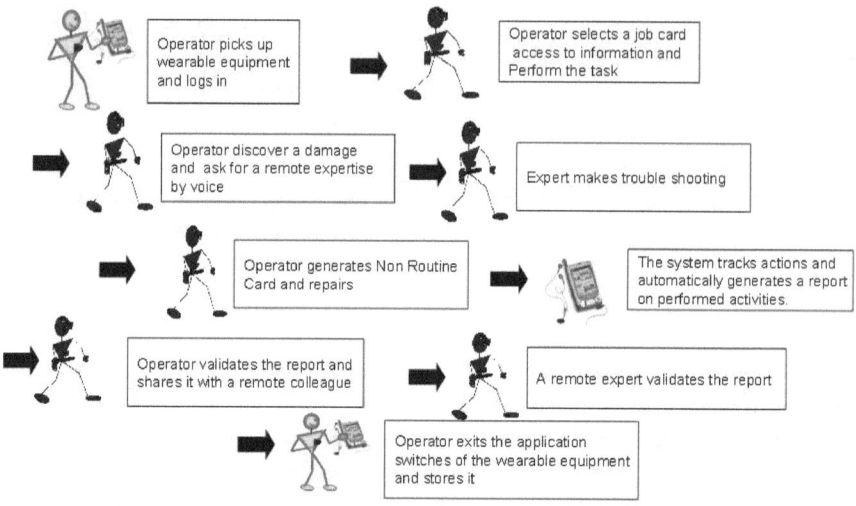

Fig. 4.3: Accessing and providing information during a maintenance task.

The challenge of gaining and accessing information through wearables, however, is how to avoid information overload. Thus, the design has to make sure that irrelevant information is screened out so as not to create a competing focus of attention. Consequently, the user needs the wearable to easily turn-off. In this regard, a central point appears to be responsibility and the ability to judge when to turn this technology off. This would be essential, as sometimes the users of a technology become the abuser of such technology, becoming its servomechanism. For instance, we have the impression that users of Smartphones often do not know when to switch them off. As McLuhan (1964) observed, there is a flip from the user using the technology to the technology using the user. This consideration points to the significance of providing training on how to make best use of this technology.

4.4 Training of end users

An important part of the use of wearables is the training component on how to use wearables effectively, mindfully, creatively, and safely. However, since the operation of most wearables is mostly straightforward there are no integrated indicators of when to use, when to ignore, what kind of problems they can create, and how to use them mindfully, hence our call for a training component. The other aspect is that with wearables, we expect the ambient support of a primary task like performing a maintenance task by providing information as a secondary task. As we know from driving a car, this requires training to learn how to split attention in a proper way. This need for training will generate new jobs and a new approach for vocational training. In this way, we use wearable technology to create apprenticeship training. As the wearable technology produces rapid change, it points to the need for new forms of training for highly skilled jobs. The old fashion ways of abstract training, classroom based or online based, will not be sufficient. What is required is on-the-job training for an effective wearable computing education.

Another training component considered when we use wearable technology is a CPS with its sensor and actuator features to support us intuitively. As a kind of training-on-the-job when users encounter an unusual situation, the wearable can direct them on how to proceed with this unusual situation. Here we can offer the examples of the "Hot wire experiment" as described in (Witt 2008). In this experiment the user has to move a tool along a shaped wire without touching it with the tool (primary task) while receiving interaction requests via a head mounted display to be answered either by speech or by gestures with the hand not using the tool (secondary task). Depending on the shape of the wire and the complexity of the interaction requests, users can be trained in different levels of expertise using the wearable technology (head mounted display and gesture) and performing mental tasks while performing different manual tasks.

Another example is the "Music glove experiment" (Huang 2010) studying Passive Haptic Learning. The authors taught the users to play piano melodies loaded into a mobile phone, which they had to play while performing other tasks (a reading comprehension test). Each note of the music played activates a vibrator on the desired finger in a specially designed glove. In a study with 16 subjects, the system was significantly more effective than a control condition where the passage was played repeatedly but the subjects' fingers were not vibrated.

Anyone using a lane-departure-warning-system in a car understands the possible achievements. In summary, wearable technology will also open up new ways of learning as long as we can use the technology intuitively.

4.5 Wearable technology in the medical sector

To further explore the considerations behind the use of wearable technology, let us describe how wearables can access information to assist in a task in a hospital (Levin-Sagi 2007). Fast access to medical information at the right moment is a crucial task for the whole healthcare system. In some cases the life of a patient can depend on it, but even in everyday routine, fast access to information means a reduction of expensive examinations and hence, cost savings. A wearable has the advantage over a tablet or a computer in that it can provide information in real time of the patient's bodily functions.

Let us assume physicians and nurses are equipped with wearable devices and adequate software systems. While on the move, during a ward-round or during a station conference medical staff also has access to all available patient information at any time and any place in the hospital. This takes into account the mobility of medical staff; provides usable, ubiquitous access to information through connections to the clinical server and audio and visual functions; speeds up data retrieval, improves the quality of the information and prevents patient mix-up through context awareness. The wearable communicates with the infrastructure like bedside displays and other devices used by colleagues, such as tablets.

The wearable technology can here improve the work of physicians and nurses in many aspects like improving the availability of information, presenting information in the actual context of the treatment situation, improving communication and knowledge sharing, and reducing the efforts for data management and documentation. Linked wearables create a knowledge network.

Medical information in hospitals is today available over electronic devices via the clinical information system that stores the information provided by physicians during examinations and therapy. Also laboratory results are delivered in large amounts to electronic medical records and can be accessed by the wearable technology for healthcare workers on the move.

The access to medical information is critical because the progress of treatment of a patient depends on the information the physician has at hand. Fast access to stored examination results can not only result in money and time savings by avoiding repeating certain exams, but also can reduce the number of unpleasant and even unhealthy examinations for a patient. It is essential that all electronic patient information is available even if spread over different information systems with limited communication capabilities among one another.

The medical treatment process is very much distributed over the whole hospital in a sense that both patients and physicians change their locations within the hospital most of the time. As a result, equipping medical staff with wearables is promising as using a laptop often requires too much attention to be used by doctors themselves. Instead, an assistant or nurse could perform this task of computer interaction or a nurse could handle paper-based information capturing requests issued by the doctor for later entry into the IT system.

4.6 Human-centered design approach

Be it in the medical sector or any other domain, there will always be a need for ICT. While technology may result in greater efficiency and boost decision-making processes, it may also compel users to mindlessly rush to judgement. We do not want that technology forces us into a subsidiary-supporting role. We want to master it. Therefore, we foster a human-centered design approach: We design our systems in a way to keep the human operator in the decision loop in control of the continuing process of action, feedback and judgment making. That keeps people attentive and engaged and promotes the kind of challenging practice that strengthens skills.

We focus on human strengths and weaknesses and thus the software plays only a secondary role. It takes over routine functions that a human operator has already mastered, issues alerts when unexpected situations arise, provides fresh information that expands the operator's perspective and counters the biases that often distort human thinking. The technology becomes the expert's partner, not the expert's replacement.

Software should be far less intrusive, giving people room to exercise their own judgment instead of acting upon algorithmically derived suggestions. Control should shift back and forth between computer and operator running an application: by keeping the operator alert while active operations become more robust. This is in some ways even less automation but much more effective automation that better matches the user's needs. We do not wish to replace human judgment with machine calculations but rather to create a decision support system that enables alternative interpretations, hypotheses, or choices. Therefore, when using a supportive context framework we always have to keep this in mind.

We need CPS and interpretive algorithms to monitor people's physical and mental states, shifting tasks, and responsibilities between human and computers. Sensing an operator struggling with a difficult procedure, the computer should take tasks to free the operator of distractions; capturing attention could be achieved in just the same way so that the CPS can actually help us and not distract us.

Ultimately, wearable computing uses body-worn devices providing helpful information relevant to a physical real-world task. Unlike mobile computing, the focus has to be on the task and the system should not distract the user from it. To fulfill this expectation wearable computing cannot depend on direct interaction with the wearer but needs to process contextual information from the user, tasks performed and the environment to infer the next required helpful action. This relevant information is the context of a wearable computing application. That makes wearable computing a more powerful tool than just a mobile computing device.

4.7 Context of wearable computing applications

There are many sources for contextual information: One way is to use the sensor features of the CPS that measure physical properties like location, lighting conditions, the presence of digital markers to name a few. Other possibilities include known aspects of the workflow and the resulting mental model of the user to take proactive actions; software generated information from databases; current time simulations; and implicit interaction by analyzing the actions of the wearer or his or her co-workers.

One problem with the wearable computing approach is the ambiguity of information gained through any system and the conversion into usable contextual information. This problem is an open research issue and many solutions for limited aspects have been proposed in the past. In what follows, we will reflect on these approaches and thus provide a state-of-the-art survey of the possibilities.

Unlike a smartphone that can save energy by turning off components until needed, a wearable computing device is virtually always on to support the user. Any software that runs on such a device needs to make a best effort at saving resources that consume a significant amount of energy – notably CPU time and wireless network transmissions. Both pervasive and wearable computing face a common problem: Information needs to be transmitted from devices in the environment to the application. There are two possible approaches: The first is creating a dedicated very efficient application from scratch, e.g. selecting the needed information providers and designing the application to directly use these sources. In the second approach, a framework is used as a layer between the application and the core framework retrieving it in an abstract way. Figure 4.4 illustrates this wearable computing framework for collaboration, communication, content management, context, location and speech recognition, user interfacing, and workflow management. While this approach obviously introduces an overhead, it makes initiating changes easy and also enables quicker development since a transport scheme is already in place. Our research in recent years was dedicated to the development of efficient frameworks for context analysis in wearable computing (Ahlrichs et al. 2014; Iben 2014; Lawo et al. 2009; Witt 2008).

A wearable computing scenario needs a fitting framework that cares equally for the information providers (sensors, actuators), and information consumers. While the consumers will typically be more sophisticated systems, their energy is limited if they are wearable. The producers might be at a fixed place in the environment and as a result have access to virtually unlimited energy but might also be a part of a mobile system or a component of the wearable system that additionally drains energy. In any case, a low computing and transmission power effort is desirable. Therefore, a wearable computing information framework has to shift

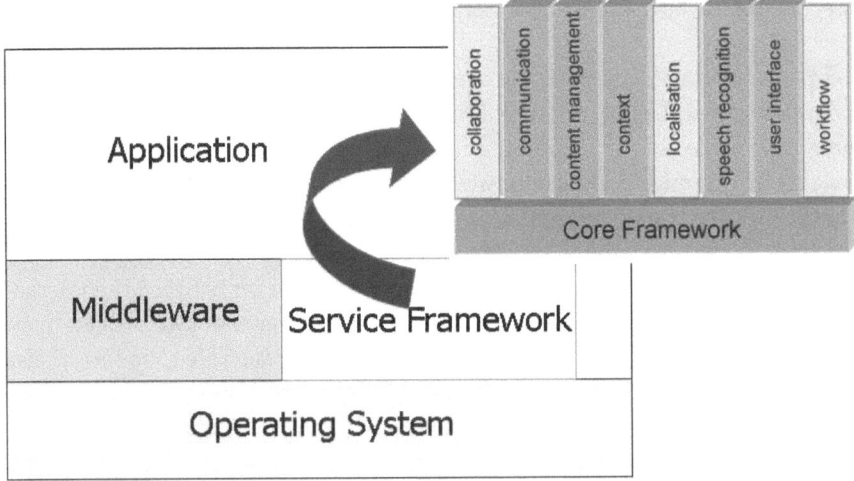

Fig. 4.4: Wearable computing framework.

needed computing power away from the individual devices and concentrate it at a suitable position. While established pervasive computing approaches create a decentralized information distribution framework, wearable computing needs a centralized approach. A single, powerful computer in the environment serves all other components. Information producers send data only to this server while it notifies consumers of the data received. This minimizes computational and transmission effort for the components by putting this burden on a single system. In addition to this, a centralized approach is also easier to integrate into a typical industrial network, as only the server has to be reachable in a controlled fashion.

Based on these concepts, the approach was implemented in recent research[1] and tested in industrial applications and in healthcare. The underlying software was published in the meantime as open source. We will introduce the approach with its benefits and results from this recent research with national and European funding.

---

[1] See e.g. www.wearitatwork.com, www.insa-projekt.de, www.rehabathome-project.eu, www. chronious.eu, www.siwear.de.

4.8 State of the art in context-aware wearable computing

Wearable computers are potentially always interacting with the environment or the user to support a task in the real world. Although wearable computing shares many similarities with mobile computing, it differs in the interactional aspect. A wearable computing application always supports a given task in the real world. That is, the user of such an application is carrying out a task while the application monitors the environment to provide needed information without disturbing the user. Attention is then the human resource that is split up between the interaction with the machine, the task, and the real world. Computer systems can try to actively seek attention from the user through various means, e.g. sound, visual cues or dialogs. The concept of multimodal interaction allows information to be interpreted in different modalities depending on the circumstances. Textual information for example can either be displayed graphically or spoken via speech synthesis. Also notifications to the user can occur via audio, by vibration or other means. In Ahlrichs et al. (2014) recent developments of frameworks for context-aware user interface design are presented.

A context-aware system incorporates information from the environment into the interaction mechanisms. This enables the system to perform actions without explicit user interaction thus reducing the amount of needed attention. context-aware interfaces are still in their early stages of development. Current commercial systems feature user interfaces that adapt to rather obvious circumstances such as device orientation or lighting conditions but are still relying on explicit input from the user to perform a requested operation. Somewhat surprisingly we are still waiting for the Smartphone that *just knows* when to switch off audible alerts during a meeting or movie although plenty of sensors are available. As our communication devices get smarter and smaller over time, the technological advances also allow the field of wearable computing to slowly reach a broader range of users. While the first body-worn devices that feature complex input and output mechanisms become available to customers, their current use is still very similar to mobile devices. What we look for are devices proactively changing their behaviour in anticipation of their user's wishes while carrying out tasks for them without needing explicit commands. One of the challenges here is not just to find a new arrangement of existing mobile user interface structures but also to embrace the available information from the environment as part of user interaction.

The thought of having devices act on their own is of course controversial for privacy and security reasons. In the end, a balance has to be found between convenience and control. This balance will have to adapt to the individual user just as the interface has to adapt to changing circumstances.

From the early days of Wearable Computing (Thorp 1998) we know that context information comes in form of events and values and that the physical source of this information is not always that important.

In Schmidt (2000) the term implicit human-computer interaction is defined and Schmidt describes there how CPS devices can be used to create context information for applications. He identified three key requirements to create software systems that can make use of this kind of interaction:

- The ability to have perception of the use, the environment, and the circumstances.
- Mechanisms to understand what the sensors see, hear and feel.
- Applications that can make use of this information.

Identifying the appropriate CPS as a sensor fulfills the first requirement. For the second and third requirements he introduces an abstraction layer allowing a uniform description of sensor events and actions taken by the application. The proposed model behind this layer consists of treating possible events as Boolean variables (too warm) determined by sensor evaluations to form complex expressions (adapt the heating). The evaluation of these expressions triggers corresponding actions in an application (the air conditioning). The concept requires that sensor readings can be transformed into an application with specific meaning and removes the need to process sensor data at the application level. Thus, smart applications do not necessary need raw sensor data (like temperature) but evaluations of it (too warm). Abstractions that assign meaning to sensor data allow replacing CPS components.

Active map services provide clients with information on located objects. In Schilit and Theimer (1994) dissemination methods for mobile hosts are investigated. The problem of transmitting information over bandwidth limited connections is analyzed for location based services for mobile applications. As a possible solution, Schilit and Theimer evaluated the idea of serving different needs of mobile receivers in terms of wanted information and bandwidth limitations. While they did not specifically create an implementation – at least not for public use – they created an elaborate architecture design to help in designing information distribution systems. The expressed ideas are very interesting from a technical point of view but the proposed systems are specialized to mainly location based information distribution. General handling of context information for more abstract information sources is not covered in this work. A core realization that also applies to general handling of information is nonetheless present: The information that clients are actually interested in is a subset of all the information that might be available to them.

4.9 Project examples

As part of the MIThril project for wearable computing at MIT, the Enchantment Whiteboard system is used for communication in general and especially for transmitting sensor information (DeVaul et al. 2003). The main purpose of the system was to accelerate the development of distributed applications. It follows the paradigm of a whiteboard where a central component can be seen from any possible client who in turn can read and write information from and to it. Additionally, clients are able to subscribe to portions of the whiteboard thus receiving automatic updates. For the transmission of high bandwidth data a secondary system exists that allows direct communication between clients negotiating a connection via the whiteboard. The whiteboard approach reduces the complexity of communication. For N clients, instead of having $N \cdot (N - 1)$ communication channels (in the worst case), a whiteboard limits communication to N as each client is only connected to a central point. The main outcomes of this research were that efficient communication is a building block for distributed applications, that automatic pushing of updated data reduces communication effort, and any framework needs to be easy to use in order to be widely adopted.

The wearIT@work project (Lukowicz et al. 2007; Lawo et al. 2011) aimed at providing a complete solution to the development of wearable computing applications with a strong focus on applications for industrial settings (see Fig. 4.5).

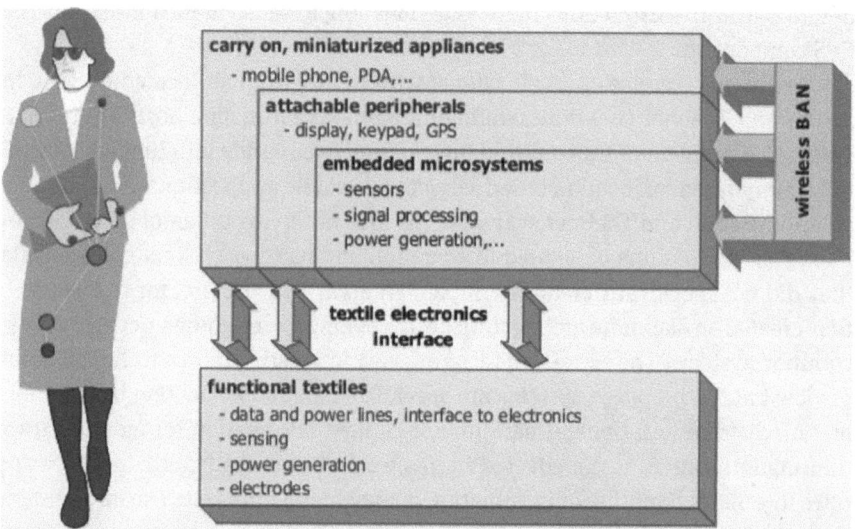

Fig. 4.5: The wearIT@work wearable computing application approach.

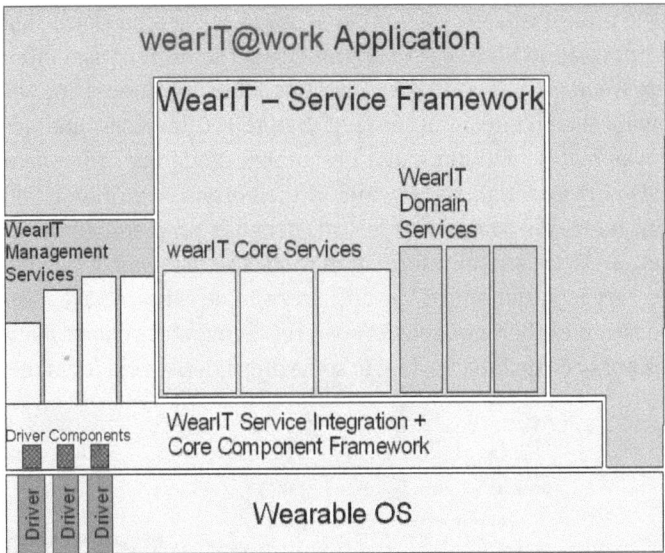

Fig. 4.6: The wearIT@work service framework.

The project followed a modular software architecture approach (see Fig. 4.6). At the centre of this approach was the Spring Framework (http://spring.io), a generic framework for modular systems using the Dependency Injection (DI) approach. This very abstract concept allows specifying dependencies between components at runtime and therefore changing components transparently to the application. The handling of context information is also implemented via DI in a way that the application does not need to know available sources of information in advance as long as the injected components are compatible. While this approach is interesting from a software engineering point, it creates an additional burden on the side of the application developer. The specific aspects of the needed context component need to be defined for all sensors or other sources of context the CPS can provide. Only with a consistent definition can different modules be injected later as replacements.

Since the meaning of information can vary greatly, this approach is not very effective in combination with DI. Furthermore, the DI approach shifts the problem of detecting available sensors into the generic framework. In the case of the Spring Framework this means that all needed sources of context have to be configured before the application starts and cannot be changed later. A component that detects sensors in the environment while the wearable application is running can of course be added as a component but would defeat the purpose of a central configuration via the framework.

A big focus of the project was the evaluation of adaptive user interfaces that can use contextual information to change their properties. Handling of user interaction is done by the Wearable UI Toolkit (WUI-TK) (for a detailed description see Witt (2008)). Following the architectural ideas of the project, the user interface specification itself was highly modular and a designated rendering component is responsible for visualizing an abstract model of the current user interaction possibilities (see Fig. 4.7). This component in turn depends on a context manager component that provides sensor information in an abstract way. While the architecture itself is very modular only a few and very simple sensors have been evaluated. A specification of the interface between rendering and context manager component has not been defined leading to some direct dependencies in this project.

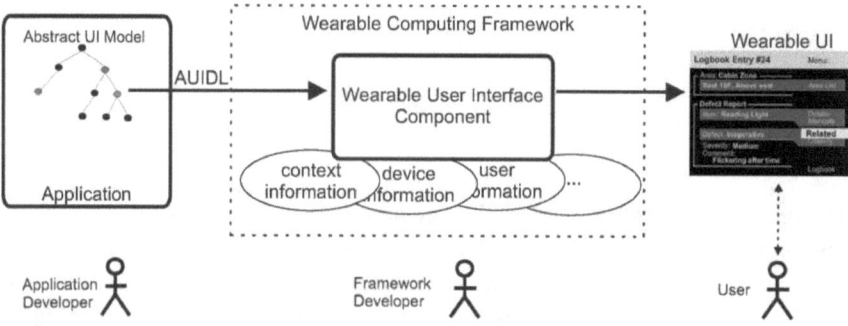

Fig. 4.7: Wearable user interface toolkit (Witt 2008).

However, the design envisions the use of distribution systems like the Context ToolKit (CTK) (Dey 2000) combined with modular systems for data acquisition. Adaptive user interfaces have to deal with lots of information to implement some kind of intelligent behaviour. Therefore, a suitable architecture for such kinds of interfaces has to include access to many different information sources (preferably in a modular way). This underlines the need for a robust way of discovering and making use of sensors in wearable applications and a separation of context handling from the logical transfer of information. A result of this research was that context information can be beneficial for a wearable system, e.g. for optimization purposes. Information is not necessarily processed at only one point in the application. Distribution systems should not make assumptions about their consumers or sources.

Bannach et al. (2006) describe a modular system for context recognition. The goal of this system is not to provide a solution for making context information available to distributed applications but to create this information easily in the first place. The Context Recognition Network (CRN) Toolbox is a set of modular components that provide access to raw sensor data and to machine learning algorithms used to extract meaning from this data. The authors of the toolbox themselves see it as a complementary tool that can be used together with a distribution system such as the above mentioned CTK. This software has been used in several projects, including the wearIT@work project. This research proved that gathering of context information can be separated from its distribution and that the separation of concerns makes adaption easier.

The Context ToolKit (CTK) (Dey 2000) is a very elaborate system that deals with creating an infrastructure for context sensing devices (CPS) and applications using the gathered information. It is engineered following an object-oriented paradigm where a common base class enables communication between objects. Subclasses are implemented for sensors, interpreters and other participants in a context-aware application. On a very low level the system relies on the HTTP protocol for communication and each object in the system acts as an individual HTTP server and client for communication. It uses a network broadcasting approach to find components in the environment, which enables context-aware applications to be used almost without configuration. However, this approach only works if the underlying network structure supports this kind of technique, which is not always the case, especially for many mobile networks. An interesting aspect of CTK is the use of the widget metaphor for working with context information. It allows the programmer to make use of a CPS (external sensor) as if it were normal type of user interface element. The widget metaphor abstracts the use of the context information from the technical means of getting the information.

4.10 Towards the TZI[2] context framework

It is obvious that communication is a key issue in designing effective wearable computing applications. When the number of external or not logically connected sources of information increases, having a framework for information distribution reduces the needed engineering effort for the application designer. The remaining question, however, is how to best structure information and how to design an appropriate communication scheme for such a framework. Many existing frameworks for context distribution have been evaluated to find common needs among

---

2 TZI – Technologie-Zentrum Informatik und Informationstechnik – (German) – Center for Computing and Communication Technology at the University of Bremen

smart applications (Dey 2000). The type of support for context in these frame-works was evaluated with respect to the type of supported information and the features provided to enable context awareness:

> There are certain types of context that are, in practice, more important than others. These are location, identity, time and activity. Location, identity, time, and activity are important context types for characterizing the situation of a particular entity. (...) Our proposed cat-egorization (...) is a list of the context-aware features that context-aware applications may support. There are three categories: 1. presentation of information and services to a user; 2. automatic execution of a service; and, 3. tagging of context to information for later re-trieval.
> (Dey 2000, p. 7)

The result of Dey's evaluation did show a high diversity in the support for these features in the various frameworks. They have been created to support a specific type of task and are not suitable for general use. This led to the creation of Dey's CTK with the goal of creating a framework that supports all needed features from the evaluated domains. While designed for general use, its main goal was to sup-port smart applications in a pervasive computing setting. In this field, constraints from the environment only have a minor impact on the application while they are much more limiting in the domain of wearable computing.

By evaluating frameworks designed with wearable computing applications in mind, more fitting criteria will be found. This evaluation will assess not only the logical needs in terms of context information but also technological needs arising from special conditions. While the outcomes from the above research show that context information can be used as an abstraction for direct input (Thorp 1998) implicit interaction in a wearable scenario is further needed (Schmidt 2000). A context information framework therefore needs to be able to process both kinds of information. Furthermore, a common problem with context information is finding an efficient way of transferring the information across many devices (Schilit & Theimer 1994; DeVaul et al. 2003). Here, a difference can be found between wearable and pervasive computing applications. Wearable applications will often encounter a change in available sources for information. These applica-tions generally need to discover and select sources according to the current needs of the user.

The wearIT@work project used a very complex interpretation of context in-formation that affects the wearable application in various ways. Not only explicit and implicit interaction has to be taken into account but also other information sources can be used to further control the wearable application, for example, by changing the representation of information to better match the current situation (Witt 2008). Finally, a general trend can be observed in favour of modular systems (Bannach et al. 2006). A context distribution system does not necessarily need means of interaction with the sources of information, rather it provides only a distribution scheme.

As a nontechnical observation, the few frameworks that have been created for solving general problems for wearable computing applications so far seem to become unmaintained (or even unavailable) after a few years for various reasons. In general, the required knowledge to operate and enhance the provided software seems to only be present among very few developers and to vanish when they leave the field of wearable computing. When a new wearable computing application is to be developed, this situation becomes the first road-block for the developers. One could argue that the availability of a framework is equally important as its suitability for the task at hand. Even if the description of a framework looks promising, a lack of resources to actually use it will encourage developers to start writing their own solutions. This is, however, true for software systems in general and not a specific problem of the wearable computing field. Applying proven methods of keeping software available to a larger community would be beneficial for new projects.

A long lasting solution for wearable computing has therefore to make use of public software distribution and documentation systems to ensure that maintainers who are not necessarily part of the original development team can take over projects easily. Consequently, the need for modular systems does not only apply to the actual software design process but should also be adopted on a higher level. There exist various independent online services for collaborative software development that are not affected by changes in working groups and have the added benefit of providing easy ways for interested third parties to contribute to a project.

The software developed at TZI by Hendrik Iben (2014) during recent years and evaluated in various projects finally became available on the GitHub software collaboration platform and can be found at: https://github.com/wearlab-uni-bremen. This popular platform allows free access to the code for any interested party. As an outcome of many publicly funded projects like wearIT@work, SiWear and Rehab@Home, a well-tested framework is now available and will hopefully be beneficial for forthcoming research for a long time.

4.11 Conclusion

The context framework discussed in the previous section supports technical solutions for our communication using wearable technologies. However, as it was argued in the introduction, what we are looking for is a Wearable Ecology (Fig. 4.1) that emerges out of the idea of media ecology of Postman and McLuhan.

Postman (1995) argues that while ICT may provide gains they also involve losses. Like Oppenheimer (2003), he draws on the recent history of technology implementation to support this view. The author claims that in the past, when technology has had positive effects, it has also carried with it several disadvantages. Often these disadvantages outweighed the advantages. After all, anyone who has studied the history of technology knows that technological change is always a Faustian bargain: "Technology giveth and technology taketh away, and not always in equal measure. A new technology sometimes creates more than it destroys. Sometimes, it destroys more than it creates. But it is never one-sided" (Postman 1990, p. 2). Postman's "Faustian bargain" argument (see O'Neill 2007) resonates with McLuhan's (1964) "extensions" and "amputations". As discussed in our introduction, McLuhan argued that all technologies are "extensions" of the body; for example the car is an "extension" of our feet. But while we seek the car for the "extension" we also receive an "amputation" in the sense that the ability of our legs to walk diminishes.

Wearable Ecology has many advantages for the end user, but there is also a price to be paid. For example, in an interview we did with a physician, he was very enthusiastic about the possibility to get information regarding medications from a wearable system but told us he is afraid of information overload and of competing messages.

One of the prices of using wearables is becoming too dependent on the technology in terms of hardware as well as software. What we described in this chapter are steps being taken in designing wearables to make sure that the user retains control of the overall situation keeping in mind that technology is never perfect in the holistic sense. Automatic speech recognition is an example of a far from perfect system. Even automobiles, which one may consider as a network of computers with four wheels and a motor fulfilling only one purpose (just bringing us from A to B) is another far from perfect system.

We can overdo automation: In case the "perfect" technology fails and no back up is there, then what? Another problem is relying too much on a given technology even though the purpose of its use is unclear. Therefore, smart systems require fault tolerance and redundancy to achieve robustness. If the implemented redundancy or fault tolerance is not sufficient, the system will lack robustness. One only realizes this when the technology fails.

But any failure challenges us to make improvements; this drives our technological development. With the Internet of Things, CPS, RFID, and Cloud Computing, we have very powerful technologies. But at what price and what do we gain?

What we try to improve with the wearable context framework is a living organism of networked molecular CPS that is not a command and control system. Any living organism automatically operates in its own best interest. It is not always the anthropomorphic CPS in control. Similarly, some processes are automatic without the need for thought like breathing, heart pumping and reflex actions. For instance, when the body overheats, one does not think "I need to sweat". Sweating is intuitive. The organism contains various kinds of CPS like reflexes or gene regulation when some genes are expressed and others suppressed.

As we argued in this chapter, when designing wearables one must consider not just the user and the function being assisted, but the ground and environment (the user's context), in which the user and the wearable operate.

> The media work us over completely. They are so pervasive in their personal, political, economic, aesthetic, psychological, moral, ethical, and social consequences that they leave no part of us untouched, unaffected, unaltered. The medium is the message. Any understanding of social and cultural change is impossible without knowledge of the way media work as **environments**. All media are extensions of some human faculty – psychic or physical ... Media, by altering the environment, evoke in us unique ratios of sense perception ... Environments are not passive wrappings, but active processes, which work us over completely, massaging the ratio of the senses and imposing their silent assumptions. But environments are invisible. Their ground-rules, pervasive structure, and overall patterns elude easy perception.
> (McLuhan & Fiore 1967, p. 26)

The transformations of technology have acquired the character of organic evolution because all technologies are extensions of our physical being. Environments work us over and remake us. It is the individual who is the *content* of and the *message* of the *media*, which are in turn extensions of the individual.

> The violence that all electric media inflict in their users is that they are instantly invaded and deprived of their physical bodies and are merged in a network of extensions of their own nervous systems. As if this were not sufficient violence or invasion of individual rights, the elimination of the physical bodies of the electric media users also deprives them of the means of relating the program experience of their private, individual selves, even as instant involvement suppresses private identity. The loss of individual and personal meaning via the electronic media ensures a corresponding and reciprocal violence from those so deprived of their identities; for violence, whether spiritual or physical, is a quest for identity and the meaningful. The less identity, the more violence.
> (Violence of the Media, Canadian Forum 1976)

Finally, any product or innovation creates both service and disservice environments, which reshape human attitudes. These service and disservice environments are always invisible until new environments have superseded them (McLuhan's Letter to Jonathan Miller on April 22, 1970 – Molinaro et al. 1987, p. 404).

4.12 Discussion and considerations for future research

We are only at the beginning of the age of wearable computing and much study and research will be required to ensure that wearable computing becomes a force that promotes human well-being and avoids the problems that often accompany the emergence of a new technology. The challenges of CPS and wearable computing in the 21st century bear a certain analogy to the challenges of nuclear technology in the 20th century. The desire for additional sources of energy to feed the needs of industrial society made nuclear technology seem like a boon to humankind until we had the experience of Three Mile Island, Chernobyl and more recently Fukushima. Furthermore, there is the fear that increasing international hostilities might result in a worldwide nuclear holocaust or today's terrorists might obtain access to a nuclear device. Now with ICT and the Internet that brought us automation, together with breakthroughs in number crunching and data manipulation, we face the new challenges of potential misuse of this technology by those that would bring harm to others as well as the possibility that users of the technology can lose their perspective and actually become enslaved by said technology. There are even those well-meaning computer enthusiasts that talk about a singularity point where computers become smarter than humans and begin to run the show. What these enthusiasts fail to remember is that humans have values while computing devices do not.

But just as we cannot dispense with nuclear energy we must also learn to live with the challenges of ICT and wearable computing in particular. Thus, we have to work on improvements to make them safer, more robust, reliable and secure. We totally support the use and development of wearable technology but as we reasoned here, we need to keep in mind that they pose dangers. For example, users can become too attached to and involved with their wearable computing and mistake the computer-generated reflection of themselves as the ultimate reality. As McLuhan (1964, p. 41) warned 50 years ago: "The youth Narcissus mistook his own reflection in the water for another person. This extension of himself by mirror numbed his perceptions until he became the servomechanism of his own extended or repeated image."

In order to deal with some of the dangers we have identified in this chapter, what is required is further research, not on how to make information and communication technology (ICT) more efficient but rather on how to ensure that ICT is used to support universal human values such as freedom, well-being for all, and the preservation of human culture. We therefore suggest the following line of study:

1. basic media competencies,
2. data security and privacy,
3. better understanding of perception and cognition and how to use this knowledge in ICT developments (wearables, robotics, human-machine interactions), and
4. media/Wearable Ecology so that all the impacts on work, education, family, culture, democracy and general well-being generated by ICT in general and wearable computing in particular are understood.

In essence, we believe that more interdisciplinary collaboration in research focused on minimizing the effect of the "Faustian bargain" of technology within the Wearable Ecology will maximize benefits and minimize any resulting social, psychological, and psychical price. The authors have been involved in such projects for many years and are optimistic that it is mission possible!

Abbreviations

CPS	Cyber Physical Systems
IoT	Internet of Things
HI	Humanistic Intelligence
IU	Information Technology
ICT	information and communication technology
CPU	Central Processing Unit
MIT	Massachusetts Institute of Technology
DI	Dependency Injection
WUI-TK	UI Toolkit
UI	User Interface
CTK	Context ToolKit
CRN	Context Recognition Network
HTTP	Hypertext Transfer Protocol
CPS	context sensing devices
TZI	Technologie-Zentrum Informatik und Informationstechnik
RFID	Radio Frequency Identification

References

Ahlrichs, C, Iben, H, & Lawo, M 2014, 'Context aware mobile and wearable device interfaces', in *Recent advances in ambient intelligence and context-aware computing*, ed K Curran, DOI: 10.4018/978-1-4666-7284-0.

Bannach, D, Kunze, KS, Lukowicz, P & Amft, O 2006, 'Distributed modular toolbox for multimodal context recognition', *ARCS*, pp. 99–113.

Clark, A, 2003, *Natural born cyborgs*. Oxford University Press, Oxford.

DeVaul, RW, Sun, M, Gips J & Pentland, A 2003, 'MIThril 2003: Applications and architecture', *ISWC*, IEEE Computer Society, pp. 4–11.

Dey, AK 2000, *Providing architectural support for building context-aware applications*. PhD thesis, Georgia Institute of Technology, Atlanta, GA, USA, AAI9994400.

Dvir, R, Pasher, E, Sekely, G& Levin, M 2005, 'Hi-tech hi-touch approach to wearable computing', *Proceedings from the 2nd International Forum on Applied Wearable Computing*, Zurich.

Forsberg, M 2003, *Why is speech recognition difficult?* Chalmers University of Technology, Göteborg/Sweden http://www.speech.kth.se/~rolf/gslt_papers/MarkusForsberg.pdf [14 January 2015].

Huang, EA, Huang K, Starner TE, Do, EYL, Weiberg G, Kohlsdorf D, Ahlrichs C, & Leibrandt R 2010, 'Mobile music touch: mobile tactile stimulation for passive learning', *Proceedings of the 28th international conference on Human factors in computing systems*, ACM, pp. 791–800.

Iben, H 2014, *Rapid prototyping infrastructure for wearable computing applications*. PhD thesis, University of Bremen.

Lee, EA 2008, *Cyber physical systems: Design challenges*, Tech. rep. UCB/EECS-2008-8, EECS Department, University of California, Berkeley.

Lawo, M, Herzog, O, Boronowsky M & Knackfuß, P 2011, 'The open wearable computing group', *IEEE Pervasive Computing*, vol. 10, no. 2, pp. 78–81.

Lawo, M, Pasher, E & Pezzio, R 2009, *Intelligent clothing: Empowering the mobile worker by wearable computing*, Aka / IOS Press, Heidelberg, Amsterdam, ISBN: 9783898386128.

Lawo, M, Pasher, E & Herzog, O 2010, 'Wearable computers and organizational change', *The International Conference on Ambient Systems Networks and Technologies*.

Levin-Sagi, M, Pasher, E, Carlsson, V, Klug, T, Ziegert, T & Zinnen, A 2007, A comprehensive human factors analysis of wearable computers supporting a hospital ward round, *Applied Wearable Computing (IFAWC), 2007 4th International Forum*.

Licklider, JCR 1960, 'Man-computer symbiosis', *IRE Trans. on Human Factors in Electronics*, March, pp. 4–11.

Lukowicz, P, Timm-Giel, A, Lawo , M & Herzog, O 2007, 'wearIT@work: Toward real-world industrial wearable computing', *IEEE Pervasive Computing*, vol. 6, no. 4, pp. 8–13.

Mann, S 2001, 'Wearable computing: Toward humanistic intelligence', *IEEE Intelligent Systems*, May/June, pp. 10–15.

McLuhan, M, Lapham, LH, 1994, *Understanding media: The extension of man*, The MIT Press, Massachusetts, ISBN 978-0-262-63159-4.

McLuhan, M, 1964, *Understanding media: The extension of man*, The MIT Press, Massachusetts, ISBN 978-0-262-63159-4.

McLuhan, M 1967, Interview, in *McLuhan, Hot and Cool: a primer for the understanding of and a critical symposium with responses by McLuhan*, ed GE Stearn, Dial, New York.

Molinaro, M, McLuhan, C & Toye W (eds) 1987, *Letters of Marshall McLuhan*, Oxford University Press, Toronto.

O'Neill, P 2007, *ICT as Political Action*. http://www.ictaspoliticalaction.com/pages/chapters/reconceptualising/postman.html

Oppenheimer, T 2003, *The flickering mind: The false promise of technology in the classroom, and how learning can be saved*. Random House, New York.

Pasher, E & Ronen, T 2011, *The complete guide to knowledge management: A strategic plan to leverage your company's intellectual capital*, Wiley, Hoboken, NJ, ISBN: 978-0-470-88129-3.

Pasher, E, Levin-Sagi, M, Dvir, R & Goldberg, M 2006, 'Using wearable computing for knowledge management: Applied wearable computing', *IFAWC, 3rd International Forum.*

Pasher, E, Popper, Z, Raz, H & Lawo, M 2010, 'wearIT@work: a wearable computing solution for knowledge-based development', *Int. J. Knowledge-Based Development*, no. 1, p. 1.

Postman, N 1990, *Informing ourselves to death*, A speech given to Gesellschaft für Informatik. Stuttgart, Germany, October 1990.

Postman, N 1993, *Technopoly: The surrender of culture to technology*, Vintage Books, New York, ISBN 978-0-679-74540-2.

Postman, N 1995, On Cyberspace, http://youtu.be/49rcVQ1vFAY

Schilit, B & Theimer, M 1994, *Disseminating active map information to mobile hosts.* IEEE Network, 8, p. 22–32, 1994.

Schmidt, 2000, Albrecht Schmidt. 'Implicit human computer interaction through context', *Personal Technologies*, vol. 4, no. 2–3, pp. 191–199.

Tennenhouse, D 2000, 'Proactive computing', *Comm. of the ACM*, vol. 43, no. 5, pp. 43–50.

Thorp, EO 1998, 'The invention of the first wearable computer', in *Proceedings of the 2nd IEEE International Symposium on Wearable Computers, ISWC '98*, IEEE Computer Society, Washington, DC, pp. 4–8.

Witt, H 2008, *User interfaces for wearable computers: development and evaluation.* PhD thesis, University of Bremen. http://d-nb.info/987814362

Ming Sun and Alexander I. Rudnicky

5 Spoken dialog systems adaptation for domains and for users

Abstract: Spoken dialog systems have been adopted across many domains and devices. However such systems are often based on a developer's understanding of the application domain and its users. But individuals differ in their understanding of a domain and in how they prefer to do tasks. Moreover the dynamics of the domain and users will change over time. A spoken dialog system that adapts to users and is aware of domain evolution would be more likely to succeed; deployed applications should take these factors into account. In this chapter we explore realistic scenarios in human-machine communication where adaptation can be used to improve the quality of interaction. We focus on two important layers within a spoken dialog system: (1) language understanding and (2) user intention learning.

The former enables the system to accommodate the user's language in its lexicon and productive aspects. The latter makes the system capable of providing complex and personalized services across multiple existing domains.

5.1 Introduction

Adaptive communication systems should adjust dynamically to the peculiarities of interacting with specific users and to reflect the structure of particular domains. With active adaptation such systems can evolve to better understand users' input and more efficiently generate behavior relevant to the needs of users. In this chapter, we focus on the problem of better modeling at two levels that users experience directly – the language they want to use and their intentions, the tasks and goals that people want the agent to help them accomplish.

Natural language understanding provides the front-end for a dialog system and its quality impacts the performance of all subsequent components in the dialog pipeline. Errors generated in speech understanding risk failure in providing utility to the user and require the user to devote time to recovery. Knowing and adapting to a user's language is an important aspect of performance.

A key aspect of maintaining good communication is being aware of a user's intentions, reflecting their goals, and their preferences in organizing their activities. These factors define the structure of a dialog. Performance can be improved by evolving task models to reflect experience and making appropriate use of an activity's context. Agents that learn about their users could also share useful information and allow a group to benefit from each other's experience.

For **language understanding**, we address the problem from two perspectives – lexicon extension and language model adaptation. In both cases, we have to face the problem of limited data. For domain applications, even with a large vocabulary speech recognizer, the vocabulary is fixed. Without the capability to dynamically expand the vocabulary (language model as well), the agent is not able to handle the evolving language of users.

We describe our work on out-of-vocabulary words (OOV) detection and recovery in Section 5.2.1. Specifically, how to proactively learn out-of-vocabulary words (OOV) from human users or online resources, given OOV examples detected by the system so far or by using the existing vocabulary. For example, when a word "table" (either within system vocabulary or not) is used, can we foresee that "desk" may be a word that the user will use later and if so, can we add it to our lexicon in advance? We describe preliminary work on (1) the potential benefit of proactive lexicon learning and (2) feasibility of using web knowledge to improve the OOV learning process.

Spoken language application developers have used either a local closed-vocabulary speech recognizer, or a cloud-based recognizer. Each has its advantages and disadvantages. For the closed domain recognizer, although the accuracy on in-domain language will be high, the coverage (lexicon-wise and language-wise) is not comparable to that of a cloud-based recognizer. Developers tend to use cloud-based recognizers but often at the cost of a drop of accuracy due to missing domain vocabulary. It is possible to configure a cloud recognizer such that language models focus on specific domains or classes of domains. However, this addresses only part of the problem, particularly for applications that should support user-specific language. A Contacts manager was an original example of this, where a user will want to speak the names that they have entered. Given a local decoder and access to local data, systems can acquire vocabulary. But more generally people will expect systems to accept a variety of specific languages, including ones that encompass oral-only vocabularies.

However, this type of adaptation, which would seem to be a reasonable way to also improve cloud ASR, is more difficult if one expects each user and device to be able to customize language. We have investigated the combination of a local domain-specific recognizer and a cloud-based recognizer that allows us to balance the advantages of each type, see Section 5.2.1. Once this architecture is adopted a number of choices present themselves, including language modeling at the local end (finite-state-grammar or statistical-language model) and the exact strategy for effective combination of hypotheses produced by each decoder. There is also the challenge of being able to train the local recognizer on very limited amounts of domain data, as would have to be expected in the case of owned systems, such as a smartphone based agent, see Section 5.2.2

Dealing with **user intentions** in the context of individual applications would appear to be tractable, especially as some applications operate in very limited functional domains (for example, a dialer dials; a music player plays music). Contemporary devices, in particular the ubiquitous smartphone, present the user with suites of applications. Each application will have limited functionality but as an ensemble they can support fairly complex activities. From the user's perspective goals and intentions operate at a more abstract level. For example, the intention may be to arrange an evening with friends but the activity will make use of several applications and will create its own context. Ideally the system can identify these structures and make use of them to better interpret what the user wants and consequently how they express themselves.

In Section 5.3.5 we describe our initial analyses of human behavior in smartphone environments, we expect that what we can learn will generalize to similar environments, such as the automobile. In Section 5.3.4, we investigate how intention information might be used to improve spoken language understanding by constraining recognition.

5.2 Language adaptation

Contents conveyed via language need to be correctly understood by a spoken dialog system and for this we need good quality speech recognition. Depending on the complexity of the domain and the expertise of developers, different approaches are available. Developers can configure their own recognizer or they can rely on cloud-based recognition. In the first approach, good language coverage is difficult to achieve, except for the simplest applications. The limiting factors will include a finite amount of training data and specifically an incomplete lexicon. A design for a usable system would need to include some process for extending a lexicon beyond that anticipated during the development cycle. The second approach can provide robust coverage of general language, but at this time cloud-based recognition will necessarily not match the specifics of the language observed in a particular domain. The consequence will be poor performance relative to a custom recognition system (Lange & Suendermann-Oeft 2014). This is particularly an issue for systems that are likely to have a significant number of idiosyncratic words, such as names in a contact list.

In this section, we discuss how such limitations can be addressed. Ideally, we want to have dialog systems that (1) learn novel words to include in the local recognizer; (2) integrate cloud-based recognition with local recognition to take advantage of better general coverage available in cloud recognition. The result is a system that evolves as language changes and maintains good performance.

We discuss the lexicon problem in local ASR in Section 5.2.1. Two major processes are involved – OOV detection and proactive new word learning. Adaptation in cloud ASR is discussed in Section 5.2.2, where an adaptation framework is described and the insufficient data problem is addressed.

5.2.1 Lexicon adaptation

Lexicons can be extended in two ways: (1) detection and capture of new words as they occur in use (detect-and-learn); (2) proactive learning of new words that may appear in a particular domain (expect-and-learn). The first approach addresses the situation when the dialog system detects out-of-vocabulary words in a user's utterance. In practice both techniques would have to be used, as unknown words by definition are unpredictable.

Detection and recovery of out-of-vocabulary (OOV) words

Speech recognition systems are primarily closed-vocabulary and will not gracefully handle unknown words. On average, one OOV word will introduce 1.2 word errors (Rosenfeld 1995) through substitution and through degradation of recognition for adjoining words. In the context of interactive systems the impact on performance is greater, as OOVs are usually open-class content words that are necessary for the task under way; good examples might be a person name or a landmark. The capability to detect OOVs and dynamically incorporate OOVs into a system's understanding model therefore becomes critical for system usability.

The out-of-vocabulary problem has been of interest for some time and various approaches to handling OOV words have been proposed. Hybrid language model with phones, subwords and graphones have been investigated (Bazzi & Glass 2000; Klakow, Rose & Aubert 1999; Bisani & Ney 2005; Schaaf 2001) and together with information such as confidence can be used to pinpoint possible OOV regions (Wessel et al. 2001; Sun et al. 2003; Lin et al. 2007). Once OOV regions in an utterance are located, phoneme-to-grapheme alignment can be used to recover the orthographic form of an OOV (Bisani & Ney 2005; Vertanen 2008).

We have investigated a fragment-hybrid language model approach to OOV word detection (Qin, Sun & Rudnicky 2011; Qin, Sun & Rudnicky 2012) and found it to be effective. We evaluated different categories of fragments in terms of detection and recovery (Qin & Rudnicky 2013a; Qin & Rudnicky 2013b). Language model score can be estimated in order to incorporate new words into existing understanding models (Qin & Rudnicky 2014). As a result, 90 % of recovered OOV words can be identified. Moreover, OOVs can be incorporated into other levels of processing such as grammars for parsing, by using either syntactic or semantic similarities, or by asking human users to provide such information (Pappu 2014).

In this approach we train an open-vocabulary word LM from a large text corpus and a closed-vocabulary fragment LM derived from the pronunciations of words in a large dictionary. Fragments can be phones, subwords, graphones, etc. The intuition is that the words in the dictionary provide syllabic and phonotactic constraints and define the expected word forms in the language; yet-unobserved words are expected to follow these constraints and should therefore be well-represented by fragments. When training the word LM, all OOV words are matched to an unknown token "⟨unk⟩". Then by combining the word LM and fragment LM, a single fragment-hybrid LM can be generated.

Phone and subword units only model the phonetic level while graphone considers orthography as well. A result of using a fragment-hybrid language model is, ideally, that in-vocabulary words are decoded as words and out-of-vocabulary words are presented as a sequence of fragments. For example, the OOV word "ashland" may show up as "AE SH AH N" in a phone-hybrid system, or "EY_SH AH_N" in a subword-hybrid system, or $\left(_{EY\_SH}^{ash}\right)\left(_{AH\_N}^{en}\right)$ in a graphone-hybrid system. Note that in these systems, OOVs may not be represented correctly via phone, subword or graphone sequences (as shown in the examples). The goal is to identify such OOV regions and recover a correct form from the possibly erroneous results.

OOV detection can be further improved by combining three individual systems together using a voting scheme such as ROVER (Fiscus 1997). The goal here is to more accurately locate OOV word in an utterance, given the OOV reported by three systems. By converting fragment sequences into an "OOV" token, three different word hypotheses can be aligned.

To recover OOV words' spellings from recognized fragment sequences, we have used a phoneme-to-grapheme conversion for the phone and subword systems. In a graphone-based system, we simply concatenate the letters together. Other methods such as using search engines (Google, Bing, etc.) to auto-correct recovered spellings could also provide more accurate spellings.

We evaluated this detect-and-learn approach using the Wall Street Journal corpus, a standard data set. We used the WSJ Nov. 92 5 k and 20 k evaluation sets (Paul & Baker 1992) and decoded using the Sphinx3 decoder. The WSJ0 text corpus was used for word language model (LM) training. The top 5 k and 20 k words in this text corpus were used as vocabulary, yielding an OOV rate of 2 % for both tasks. Then an open-vocabulary 5 k-word LM and 20 k-word LM were trained. The dictionary was generated using CMUDict (v.0.7a). The fragment LM was trained from the dictionary. We used the WSJ-SI284 corpus to train our acoustic model and we trained bigram fragment-hybrid models for phone, subword and graphone. Word-error-rate (WER) using the word bigram LM was 9.23 % for 5 k task and 12.21 % for 20 k task.

We report recall and precision of OOV detection to evaluate an individual system's performance (see definition below). Recall and precision is calculated at the word level which measures both the occurrences and positions of OOV words in an utterance, since for practical purposes knowing where OOVs are located in an utterance is more valuable than simply knowing that an OOV has occurred. Location information can be used later when communicating with human users about OOVs and their meanings, using the context provided. For example if the decoding is "My name is ⟨OOV⟩" it should be possible to then reply something like "Your name is ⟨OOV⟩. Is that right?". To evaluate OOV recovery, we can compare word error rate before and after incorporating the written form of an OOV word into the vocabulary.

Figure 5.1 shows that the subword approach outperforms other hybrid systems in the 5 k task. Both subword and graphone system are able to use longer fragment history than a phone system. But a graphone system has too many variations, even if we constrain the graphone length to be as short as 2, which requires more training data for reliable language model estimation. For the 20 k task, the graphone system performance catches up with the subword system and they are both better than phone system.

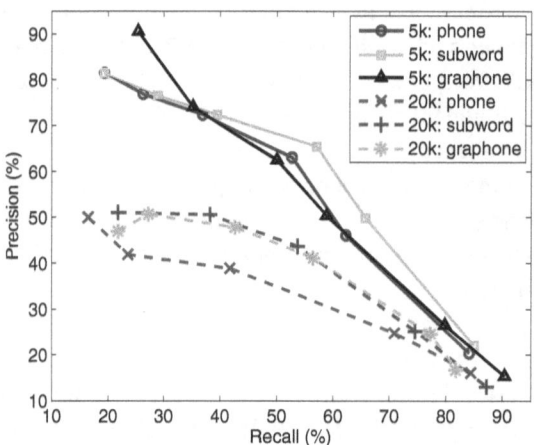

Fig. 5.1: OOV detection performance for different fragments.

Figures 5.2 and 5.3 show OOV recovery performance. Notice that the baseline recognition error without OOV detection and recovery is 9.23 % and 12.21 % respectively. From these figures, we can see that: (1) OOV recovery improves recognition (compare each solid line with dashed line); (2) subword hybrid system is the best in 5 k task and graphone is the best in 20 k task (similar to the detection task); (3) in the low recall region, system WER can be lower than baseline, which implies that OOVs can be detected with hybrid models without affecting recognition performance.

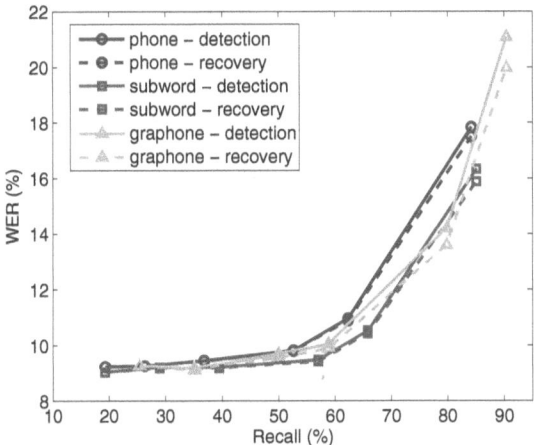

Fig. 5.2: OOV recovery results on 5 k task.

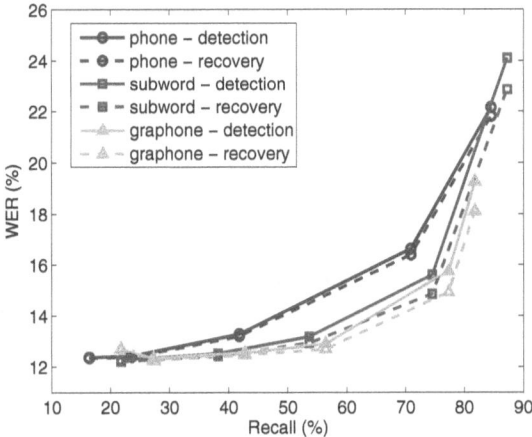

Fig. 5.3: OOV recovery results on 20 k task.

Summary. We show that OOV detection and recovery are feasible using a fragment-hybrid language model. Among three different fragment-hybrid systems, we find that subword and graphone hybrid approaches are better than a phone hybrid system for both OOV detection and recovery. This is likely due to the constraints afforded by larger units that reflect the lexical structure in a language.

Proactively learning OOV words

From the previous section, we know that OOV words can be detected by using a hybrid language model during decoding and their orthographic forms can be recovered by applying a phoneme-to-grapheme conversion. While important for capturing novel words (such as names) interactive systems nevertheless should minimize interruptions to the flow of a conversation. This should happen from the start, by predicting yet unobserved words given the initial dictionary. While some words, such as people names, may not be directly predictable the language that people will use will still be communicating concepts relevant to the application.

We now describe the potential improvement to spoken dialog system performance in speech recognition and semantic understanding if semantically relevant vocabulary can be acquired at the outset. In our experiment, we used SEMAFOR, a state-of-the-art frame-semantic parser, to produce semantic representations from recognized speech (Das 2012). SEMAFOR parses consist of semantic frames that encode events, relations, or entities and the participants in the utterance (Baker et al. 1998).

We use the parses obtained from manual transcriptions as the gold standard semantic representations. We report the recall and precision of semantic frames (including words that evoke those frames) by comparing the parses from speech hypotheses with reference parses.

To establish an upper bound we can compare recognition performance for three different cases: (1) large generic vocabulary; (2) the training corpus vocabulary; (3) training vocabulary + potential novel words in the testing set which can be found in the large generic dictionary. The OOV rate, WER and speed are shown in Tab. 5.1. Note that for WSJ or SWB, acoustic model and language model are the same across three types of vocabulary. For each data set, we report word-error-rate (WER), recognition speed (times real-time, xRT) and the recall and precision of the decoded frames for understanding, as described earlier. From Tab. 5.1 we can see that, by simply adjusting which words can be activated during decoding, both recognition and understanding performance can be improved. Note that in both WSJ and SWB, when potential OOVs are added to the domain vocabulary, error rate drops below the Domain condition, with only modest increases in vo-

Tab. 5.1: System performance for different vocabulary composition and sizes. Speed is shown for reference; no parameter tuning was done.

Data set	Vocab	Vocab Size	OOV Rate	Recognition		Understanding	
				WER	xRT	Recall	Precision
WSJ	Generic Vocab	19792	4.6	22.01	0.09	84.7	79.9
	Domain Vocab	2854	22.6	49.90	0.05	62.6	52.3
	Domain + Recoverable OOVs	3997	**3.6**	**20.44**	0.05	**85.9**	**80.3**
SWB	Generic Vocab	19792	1.8	43.87	1.17	64.8	62.2
	Domain Vocab	1218	13.2	54.61	0.88	54.6	50.9
	Domain + Recoverable OOVs	1778	**1.6**	**42.14**	0.90	**67.5**	**63.8**

cabulary size and decoding time. Our goal should therefore be to fill in the gap between existing training vocabulary and working vocabulary.

Automatically adding pronunciation variations from a phoneme decoder to an existing word dictionary (Sloboda & Waibel 1996; Strik & Cucchiarini 1999) has been investigated by others. In our work, we primarily focus on proposing new words rather than learning pronunciations. We therefore use pronunciations taken from the standard CMUdict dictionary (CMU Pronouncing Dictionary 2015). Paraphrasing has been studied to capture lexical variations in small domain (Prud'hommeaux & Roark 2012). Their application was to find predefined important elements from multiple versions of storytelling. Online resources can be used for finding similar words or phrases in a large data set (Miller 1995). Words can be transformed into feature vectors and similarities can be computed among such vectors (Mikolov, Chen et al. 2013; Mikolov, Sutskever et al. 2013). We focus on learning new words to improve language coverage for reducing WER.

Experiment. We investigated web-based OOV learning using two well-studied corpora, read speech from the Wall Street Journal (WSJ) and conversational speech from Switchboard (SWB). We made use of offline resources which provide information about word similariry, knowledge-engineered (WordNet) and corpus-based (Word2Vec).

We investigated the quality of web-based learning strategy in the testing set in terms of recall of the potential OOVs (defined in equation (5.1)) and the number of proposed new words. In the equation below, correctly proposed OOVs are those OOVs learned from web or human that actually occur in the test set.

$$\text{Recall} = \frac{\text{\# correctly proposed OOVs}}{\text{\# OOVs in reference}} \times 100\,\% \tag{5.1}$$

Web-based OOV learning

We used Word2Vec (Word2Vec n.d.; Mikolov et al. 2013a; Mikolov et al. 2013b) to find closely related new words given existing vocabulary. We tested on Wall Street Journal (WSJ) and Switchboard (SWB) data sets. For WSJ, training and testing sets both contain 546 utterances. For SWB, training contains 546 utterances and testing contains 545. We have two development sets, each containing 300 utterances for WSJ and SWB respectively.

We compared our idea of using similar words to randomly selecting new words. We have two baselines: (1) randomly select a word from a given word's similar words; (2) randomly select a word from the generic dictionary as mentioned earlier. Figures 5.4 and 5.5 compare selecting most similar words with the two baselines above. The system is always using the most frequent words in training data as seeds to learn new words. We can see that randomly selecting words to incorporate into vocabulary (baseline1 and baseline2 in dashed lines) is not as good as using the closest words (solid lines) for both data sets. In this section, we present our thoughts regarding how to effectively add new related vocabulary into the existing models.

We had two intuitions regarding the nature of OOV words: (1) OOV words are more related to frequent words in vocabulary (e.g. "table" is frequently used so "desk" may be used as well); (2) OOV words may happen to be infrequent words in vocabulary since statistical understanding models capture frequent phenomena in general. From Fig. 5.5, we can find that for the SWB data set, using most similar words of a given vocabulary reaches best performance compared with two

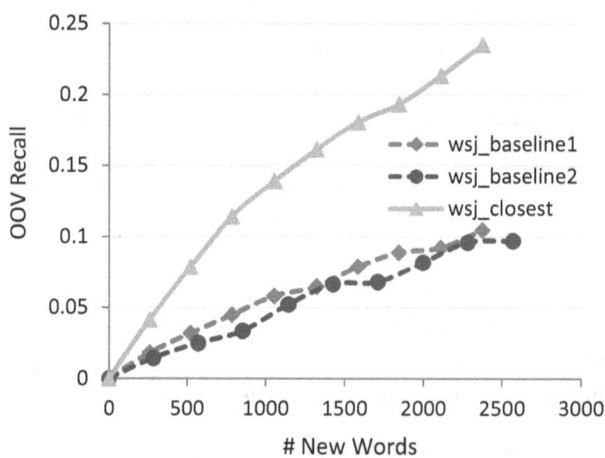

Fig. 5.4: Closest words vs. random words (WSJ).

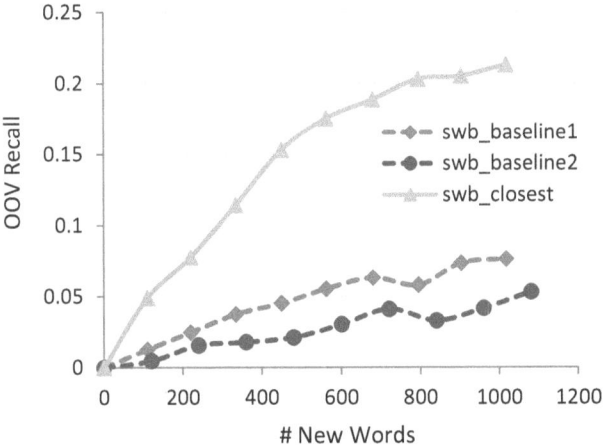

Fig. 5.5: Closest words vs. random words (SWB).

baselines. However, the performance increase slows down as it reaches infrequent words as the abscissa is ordered by word frequency. For the WSJ data set, it is not very clear. We conducted a comparison by expanding frequent words or infrequent words for WSJ data set. The token level OOV recall on the test set is shown in Fig. 5.6. The solid lines denote only one new word is learned for each seed. The dashed lines denote that two words can be learned for each seed. The "reverse_wfreq" indicates that the least frequent vocabularies are first used as seeds. From the figure we can see that using frequent words as seeds is better than using infrequent words in terms of OOV recall given the same amount of new words being learned.

Second, we want to understand how many new words per in-vocabulary seed generates. As we can see from Fig. 5.7, the quality of acquired new words degrades as more new words can be allowed per seed. This is intuitively reasonable since we are always introducing the most similar words into vocabulary.

The results above are purely based on training vocabulary. Now the question is what if we have some understanding of what potential OOVs look like, e.g. some OOVs may be detected and recovered. As described earlier, we have a development set of 300 utterances. Assuming we have detected the OOVs in the development set or knowing the part-of-speech distribution of development OOVs, we investigate the quality of new vocabulary that can be acquired with and without this knowledge. Part-of-speech distribution of OOVs in WSJ development set is shown in Tab. 5.2. As shown in Fig. 5.8, the quality of expanding frequent OOVs in the development set or the frequent words in training vocabulary are very similar. But

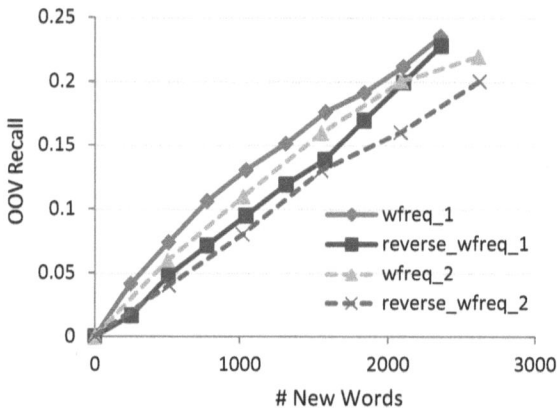

Fig. 5.6: Compare frequent and infrequent Vocab as seeds (WSJ).

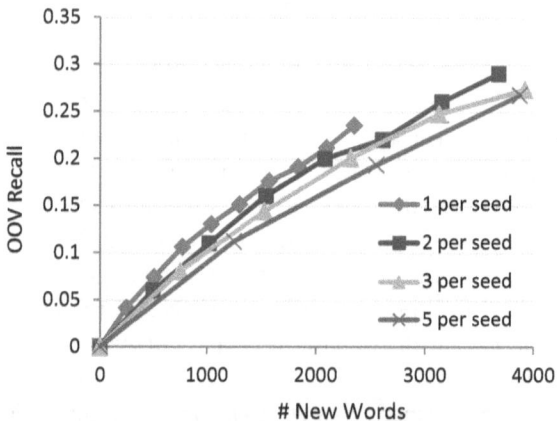

Fig. 5.7: Vary # of new words per seed (WSJ).

Tab. 5.2: Ranked OOV POS tags (WSJ).

Rank	POS	Percentage
1	NN	38.1
2	NNS	15.8
3	JJ	11.1
4	VBG	5.5

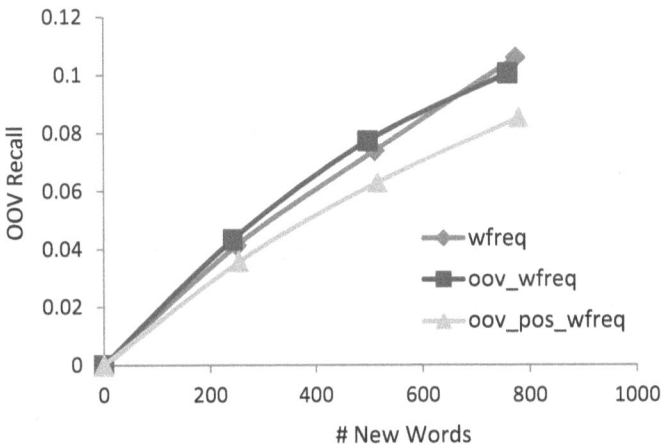

Fig. 5.8: With or without OOV knowledge (WSJ).

both are better than expanding training vocabulary using the POS tags of frequent OOVs. This indicates that information about potential OOVs may not be readily usable. More investigation is necessary to incorporate such information.

To investigate the overall recognition performance, we adopted a general language model – the Sphinx US Eng Generic Trigram Model (Sourceforge 2015). We interpolated this language model (with weight 0.7) with a language model trained on the training set (same size as the testing set). We used Wall Street Journal acoustic model. One new word per seed is allowed for each vocabulary in the training set. The new word has to be close to the seed and it has to be in the generic language model. We report WER as number of new words increases. Figure 5.9 shows that the error rate decreases as more new words are proposed.

From these preliminary experiments, we conclude that (1) selecting important seeds is useful for finding new words; and (2) new words can be found given the seeds. We propose that further improvement to the new words discovery process is possible by (1) building a classifier to find useful seeds and (2) ranking similar words appropriately so that potential OOVs can be ranked higher and can be included.

Summary
We proposed that recognition systems can proactively learn new words based on existing vocabulary or detected OOVs. We evaluated web-based strategy in terms of the recall of proposed OOVs and recognition performance to understand the characteristics in the field of proactive lexicon learning.

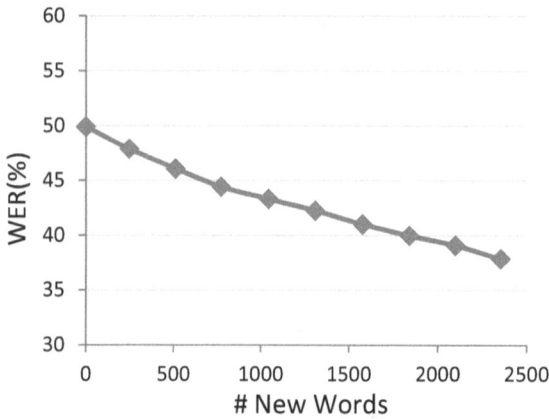

Fig. 5.9: WER as more new words incorporated (WSJ).

5.2.2 Adapting cloud ASR for domain and users

Cloud-based recognizers perform robustly across different domains by virtue of large training data sets and very large vocabularies. However, such recognizers are difficult to adapt for specific applications or users because the language model is generic. We propose that the solution should be an adaptation framework that uses local, adaptive recognizers in combination with a cloud-based service. The advantage of doing so is to exploit the domain/user adaptability of the local ASR while using the large coverage provided by cloud ASR. This combination can potentially compensate for the weakness of the respective recognizers. The framework is discussed in Section 5.2.2.

In practical use cases, spoken language system developers regularly face the problem of limited domain data. We are interested in domain/user adaptation given such a constraint. We propose to address this problem in two aspects – quality and quantity. Specifically by incorporating limited domain/situated knowledge.

Domain and user adaptation framework

Large-vocabulary cloud-based speech recognizers are widely used because of their stable performance across environments, domains and users. For example, Google ASR (Schalkwyk et al. 2010) is commonly used in Android application development. However, specific domains or users will have their own language behavior which will not be precisely captured by cloud ASR (Morbini et al. 2013). Intuitively, adaptation to domains, users or environment should improve system performance (Leggetter & Woodland 1995; Bellegarda 2004; Wang, Schultz & Waibel 2003). In a common use case, smartphones can collect personalized data and use it for adaptation. As of this writing, such adaptation data do not appear to be practically communicable to a cloud-based ASR; even then, privacy issues would have to addressed. In our framework, we use collected domain/user dependent data to directly build a local language model. A cloud-based ASR is used in combination with the local ASR to improve overall performance.

Related Work

Previous research has examined combining domain-/user-independent Google ASR with domain knowledge to improve the recognition by filtering Google results with domain constraints at word level or phonetic level (Milette & Stroud 2012; Twiefel et al. 2014). This approach post-processes Google recognition hypotheses but requires a defined domain and assumes a set of restrictions. Anything beyond the defined domain is not allowed. In addition to the difficulty of precisely defining a domain language, out-of-domain sentences will still be encountered, especially when switching among voice applications via speech. Our proposed method adapts to the domain/user dynamically and out-of-domain hypotheses are not excluded.

In speech recognition, the language model constrains the set of utterances that can be recognized. Two statistically-based paradigms have traditionally been used to derive prior probabilities of a word sequence: finite state grammars (FSG) and n-gram models (SLM) (Mohri, Pereira & Riley 2002; Jelinek 1991; Bellegarda 2000). Rule-based finite state grammars are reported to perform better than n-gram models on in-grammar sentences but perform considerably worse on out-of-grammar utterances (Knight et al. 2001). Integration of these two models has been previously investigated and has been shown to overcome individual model's weakness (Bellegarda 2000; Meteer & Rohlicek 1993). In this work, we train probabilistic FSG and SLM models from the same data and compare their performance given language model adaptation under different circumstances.

Model interpolation can be used for adaptation (Bellegarda 2004) by training separate models (an adapted one and a background one) and combining the two models. Alternately, we can merge the adaptation corpus and background corpus together and train a single model. In today's use cases, such background model/ corpus is expensive and time-consuming to create and run locally on smart devices. Cache-based language model adaptation has also been used (Jelinek et al. 1991). In dialog systems, when context is available, dynamic model switching can improve recognition performance (Xu & Rudnicky 2000). However, this method works when the cache size is sufficiently large – implying that a long discourse history is needed. In real-life situations, especially in smart device voice applications, interactions with a given application will usually be short in length. Similar to the model interpolation, this approach still requires running a large vocabulary recognizer locally on the device.

Local recognizer comparison

To compare FSG and trigram SLM performance we ran 10-fold cross-validation on 70 % of a total 1761 utterances to compare FSG and trigram SLM performance. In each fold, 90 % of the data is used to train FSG and SLM models and the rest 10 % is used for validation. Evaluation is performed on the whole training set (70 % of total utterances) and the validation set (10 % of training examples) respectively in each fold. For the entire training set, on average the out-of-language rate (OOL rate: how often a word sequence is not seen in the training data for models) is 3 % while for validation set out-of-language rate is 34 %.

From Fig. 5.10 (a) and 5.10 (b) (averaged WER over 10 folds) we see that first, for in-language data, both FSG and trigram SLM perform significantly better than cloud-based ASR ($p < 0.01$) and FSG is better than SLM (significantly in the training set); second, for out-of-language sentences, the domain-specific trigram model is better than cloud-based one and they both are significantly better than FSG ($p < 0.01$). These observations indicate that choosing an appropriate local language model may depend on the expected out-of-language rate. This is consistent with the overall performance in our experiment – when most utterances are observed (as in the training set in Fig. 5.10 (a)), both FSG and trigram significantly outperform cloud-based ASR ($p < 0.01$) and FSG is significantly better than trigram; when fewer sentences are used for training (as in the validation set in Fig. 5.10 (b)), trigram significantly outperforms the other two ($p < 0.01$).

We conclude that, in a small domain, domain/user dependent local language models will improve speech recognition accuracy compared to cloud-based recognition (this should not be surprising). FSG is more suitable when OOL rate is low. Otherwise trigram is better suited.

Combining local ASR with cloud ASR

Local recognizers can improve system performance significantly when adapted to the domain and users. But by examining the errors, we can see that local FSG produces many errors in out-of-language utterances, whereas cloud ASR does not. For a local trigram model, although its performance on OOL utterances is better than cloud ASR, it has difficulty handling utterances that are either rare in training or include out-of-vocabulary words. For example, in "search for Dorothy Green",

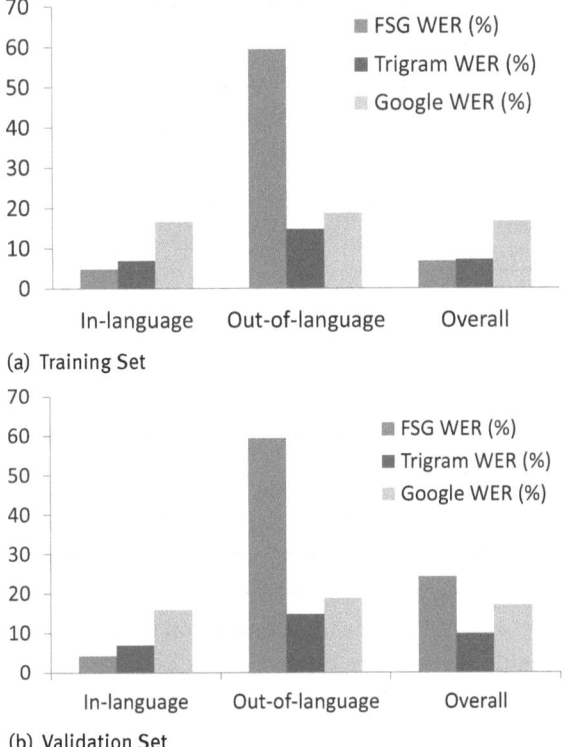

(a) Training Set

(b) Validation Set

Fig. 5.10: 10-fold Cross-validation on Individual System's Performance (WER).

the word "search" is of low probability in language model and the word "for" is not seen in training at all. Therefore, combining the local recognizer with the cloud one is intuitively promising.

In this experiment, we combine local recognition with cloud-based recognition to quantify the benefit. Our goals are: (1) to gauge the potential of system combination if correct selection can be made between local ASR and cloud-based hypotheses (an oracle baseline); (2) to build a classifier, using readily available features generated in recognition to pick the most likely correct hypothesis.

To train such a classifier, we take 70 % of our data for training (using the remaining 30 % as a test set) and further divide the training into 50 % and 20 % portions. Local FSG and trigram models are trained on the 50 % part. We use FSG, trigram model and cloud ASR respectively. Individual system performance is shown in Tab. 5.3 (WER is shown next to system names). Since the FSG and trigram SLM training corpus is a portion of the training data, there are out-of-language utterances for these two models in the training and testing sets of the classification model. The OOL rate is 11 % for training set and 38 % for testing set. Among individual systems, the trigram adaptive system performs the best. Its WER is less than 45 % of cloud WER in training and less than 60 % of cloud WER in testing. FSG performs poorly on testing data, due to the high OOL rate. Cloud ASR does not show much difference across training and testing set (as would be expected).

Tab. 5.3: Ranked performance of individual systems.

Rank	Train WER (OOL: 11 %)		Test WER (OOL: 38 %)	
1	Trigram	(7.4)	Trigram	(10.1)
2	FSG	(11.0)	Google	(17.6)
3	Google	(16.6)	FSG	(27.2)

We evaluated combinations of the three individual systems to determine potential improvement. Three two-system combinations (FSG + Google, Trigram + Google, FSG + Trigram) and one three-system combination (FSG + Trigram + Google) are considered. In a real-life use case, depending on the availability of computing power and network connectivity, some individual or combination of the individual systems can be selected according to circumstance.

To combine individual systems, those cases in which recognizers do not agree with each other are labeled as *fsg*, *trigram* or *cloud* depending on which recognizer produces the lowest WER compared with the reference transcription. We use logistic regression from SKLEARN (Pedregosa et al. 2011) to train a classifier for two-system combination. For combining three individual systems together, we use K-NEAREST. The classifiers are based on basic features: (1) the number of words in each system's hypothesis; (2) utterance acoustic score (accessible from FSG and SLM systems though not from cloud ASR). We show oracle WER on training and testing sets in Tab. 5.4. We also report WER achieved using the basic classifier in Tab. 5.5. In both tables, systems are ranked according to WER (shown next to system name). Systems which outperform all the individual systems within the combination are italicized.

Tab. 5.4: Ranked oracle WER performance for different system combinations.

Rank	Train (OOL: 11 %)		Test (OOL: 38 %)	
1	*FSG + Trigram + G*	(2.3)	*FSG + Trigram + G*	(4.4)
2	*FSG + Trigram*	(3.3)	*Trigram + G*	(6.0)
3	*FSG + G*	(3.6)	*FSG + Trigram*	(6.4)
4	*Trigram + G*	(4.5)	*FSG + G*	(8.6)

By comparing the oracle results of system combination (Tab. 5.4) and individual system's performance (Tab. 5.3), we can see that potentially, system combination will outperform any individual system. Combining local FSG, trigram with cloud-based (G in both tables) recognition provides the best potential recognition accuracy. From Tab. 5.4 we can also see: (1) combining more information sources (recognizers) improves the performance; (2) in training, where the out-of-language rate is low, the performance for in-language utterances is most important. Therefore, combinations with local FSG as one component is better than those without FSG; (3) in testing, where in-language utterances and out-of-language utterances are more balanced, the adapted trigram demonstrates its advantage in handling both cases (as described in Section 5.2.2). Combinations including the trigram model should therefore be preferred.

Tab. 5.5: Ranked WER performance for system combinations, using basic features.

Rank	Train (OOL: 11 %)		Test (OOL:38 %)	
1	*FSG + Trigram + G*	(3.4)	*Trigram + G*	(9.5)
2	*FSG + Trigram*	(5.0)	*FSG + Trigram + G*	(11.1)
3	*Trigram + G*	(6.5)	*FSG + Trigram*	(11.9)
4	*FSG + G*	(6.9)	*FSG + G*	(15.5)

Practically, as shown in Tab. 5.5, in both training and testing all system combinations achieve lower WER than cloud ASR by itself. For the training set, all combinations beat individual systems in the combination. For test set, all combinations show at least 12 % relative improvement from cloud ASR. Trigram combined with cloud ASR shows 5 % relative improvement compared with trigram alone and 46 % compared with cloud ASR.

To summarize, we found that system combination has substantial potential for reducing WER, compared with single systems. Combining local ASR (FSG or trigram) with cloud-based ASR can practically (without requiring sophisticated feature engineering) be used to adapt systems to specific domains or speakers while maintaining the benefit of a large-vocabulary cloud-based recognizer.

Required data for adaptation

After observing the benefits of combining local ASR with cloud-based ASR as an adaptation approach, the next question we want to address is how the amount of adaptation data governs improvement. In this study, we vary the amount of adaptation data available for FSG and trigram models and note performance. The ratio between the size of training set and the size of testing set is from 0.5 : 1 to 3 : 1, reducing out-of-language rate from 73 % to 29 % in testing set. Recognition performance on the test set (439 utterances) is reported. Similar to the previous section, we also report the potential improvement of combining individual systems using an oracle reference. Results are shown in Fig. 5.11.

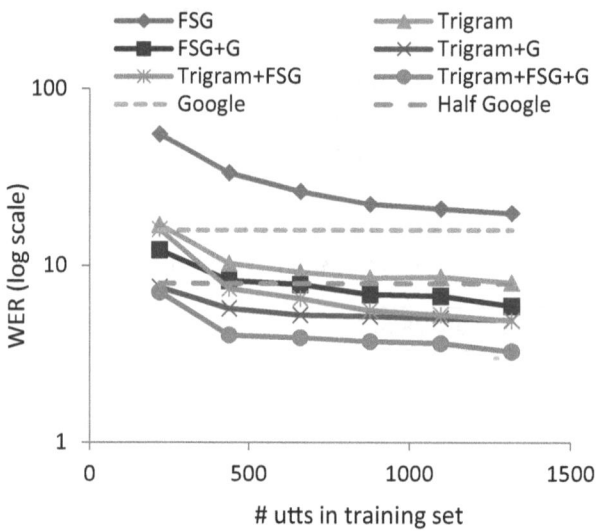

Fig. 5.11: System performance (log WER) on different amount of adaptation data.

We can observe that when more data are available, individual local systems (dashed lines with markers) perform better and oracle performance of system combinations also improves (solid lines). The two horizontal dashed lines without markers are WER for Google ASR (higher one) and half of its WER as reference baselines. As we can see, for individual systems, FSG WER is approaching Google performance as more training data becomes available. On the other hand, the trigram model beats cloud WER even with a small amount of training data (a little more than half of the size of testing data). Moreover, as additional training data is accumulated, trigram performance actually reaches half of the cloud WER.

Combining trigram with cloud ASR (noted as G in the figure) improves faster than the other two combination (FSG + cloud and FSG + trigram). When more data is added, it can reach as low as 30 % of cloud WER. With training data whose size is only half of the testing data, combining trigram with cloud ASR can potentially achieve half of cloud WER, while the other two combinations require either equal size of the testing data (FSG + trigram) or more (FSG + cloud) to potentially achieve the same performance. Combining three systems together is potentially even more useful, reaching as low as 20 % of Google WER.

Summary

A real-life limitation of cloud-based speech recognition is the drop in performance observed in application domains that introduce specific vocabulary (for example, personal names) and user language-use patterns. We investigated a potential solution to this problem that involves supplementing cloud recognition with specialized recognizers that are directly adapted to the domain/speaker. We find that rather modest amounts of adaptation data are sufficient to produce better performance. This suggests that in some use cases, in particular personal devices with limited computation, a combination of specialized recognition residing on the device, combined with cloud recognition (when available) may show superior performance in comparison to any single recognizer. We should note that our hypothesis-selection classifier, even in its simple form, produces better performance. We expect a more systematic investigation of combination strategies will yield even better performance: for example, word-level combination can produce better performance (as we have observed in other domains).

Adaptation with limited data

We always face the problem of limited domain data available for training when building a spoken dialog system, especially for speech understanding models. Where are the initial data to come from? We face two constraints in this issue – **quality** and **quantity**. With respect to quality, we need to address unsupervised adaptation which takes some user utterances and obtains reliable transcriptions with limited sources of information. For quantity, we studied how to incorporate some, but limited, domain knowledge to generate a sufficient number of training examples.

Improving quality of adaptation data

To handle the quality issue, a smart device needs to take advantage of a cloud service to obtain reliable transcriptions from user speech. Two possible obstacles are (1) user speech is very situated (such as the contact list in the phone), which may fail cloud ASR and (2) cloud ASR may naturally fail for some utterances (e.g. poor WER). To handle problem 1, we can make use of some domain knowledge such as contact list, installed application names, etc. to recover true information in cloud ASR hypotheses. To solve problem 2, the system needs to discard poor hypotheses using, for example, dialog states (if the user says "No you are wrong!" then remove the hypothesis). We focus on recovering reliable information from user speech via cloud ASR.

In the initial deployment of a speech system, labels for accumulated observations such as user's speech input are expensive to obtain. An unsupervised method is preferred (Bacchiani & Roark 2003; Gretter & Riccardi 2001; Stolcke 2001). However, this always implies that an unknown service decodes utterances and sends back the transcriptions. Confidence score can be used to prune out or discount unreliable hypotheses in acoustic model adaptation (Anastasakos & Balakrishnan 1998) and language model (Gretter & Riccardi 2001). But, in this fashion, some otherwise useful sentences may be lost – reducing the amount of training data which is already sparse. Or wrong information will still be there although discounted. It has been proposed to use phone-level information from hypotheses, together with domain knowledge to recover error-prone hypotheses into reliable ones (Twiefel et al. 2014). We speculate that these two methods can be integrated such that a score can be assigned to a recovered hypothesis depending on how many elements have been altered and whether the change is major. Thresholds can be used to discard poorly recovered ones.

Handling limited amounts of adaptation data

For quantity, there are two ways to address the problem: (1) build speech recognition based on abstract domain knowledge such as grammars handcrafted by developers; (2) collect sufficient speech data to build a statistical-language model with its corresponding vocabulary. The first technique relies on human knowledge, understanding how people would act in a given domain. The second one requires a large training data set.

Language model adaptation requires a relatively lower amount of adaptation data compared with the background model estimation (Bellegarda 2004). However, at the initial phase of system deployment, data is difficult to obtain. Without adaptation, the system may fail the user and this initial experience influences the drop rate of speech applications. Not to mention that a general adaptation framework which includes one small model and a large background model may not be feasible in real-life environments such as smartphones. An adaptation framework combining two decoders might be the best choice (as described in Section 5.2.2). Thus, accumulated user utterances are used to train the local model. No matter which framework we adopt, the limited amount of training/adaptation data is a key problem. Researchers have worked on language model training with small corpora (Placeway et al. 1993). Words can be grouped into semantic classes. Alternatively, using a similar idea, we can sample a grammar (with semantically grouped hierarchical classes) to build an artificial corpus for language model estimation, suitable for spoken dialog systems where domain corpus is expensive to gather (Kellner 1998). Other fields such as plan recognition also use this method (Blaylock & Allen 2005). In Olympus dialog systems, the Logios component adopts a similar approach to sample the grammar and create a pseudo-corpus for LM training.

5.2.3 Summary

We investigated two aspects of adapting recognition models – lexicon and language patterns. We built a framework to learn out-of-vocabulary words when such are detected. We also investigated OOV detection and recovery performance on different types of hybrid systems. The experimental results show that there is a potential for using web knowledge to learn new words. We propose that proactively learning OOVs using web knowledge or human knowledge is a key part of the solution to this challenge.

We designed an adaptation framework to make cloud ASR suitable for specific domains or users. We found that accurate recognition of in-domain language by a local recognizer can be combined with the coverage of cloud ASR to produce better performance than either alone. We also proposed that dealing with the quality issue of initial data for building such an adaptive system could be approached by recovering reliable information from cloud ASR hypotheses.

5.3 Intention adaptation

5.3.1 Motivation

With many applications installed on their devices and various tasks on their mind, users have different ways to perform particular tasks using different applications. For example, both user A and B want to organize a dinner with their friends. User A launches Yelp to find a good restaurant and called his friends with the address and time to meet up. User B calls a friend to ask for a recommendation and messages her friends about the details and then calls the restaurant to reserve a table. Reasonably, people have their own preference regarding organizing the steps to achieve a certain goal. We investigate such various intentions and the way intention is organized. There are two goals related to user intentions: (1) recognize the current intention of the user; (2) use the identified intention to improve communication.

Possible impacts on performance would be: (a) prepare the dialog models (e.g. recognition, dialog task, etc.) given a certain intention; (b) evaluate current user's application usage and recommend a more efficient path. For example, if we know user A wants to schedule a project meeting with classmates, the system may anticipate the user A's language to include communication (e.g. "Ask them if 6 pm is good to meet") and calendar operation (e.g., "Set alarm for 5:55 pm tomorrow for group discussion"), etc. If we know user B spent 30 minutes on a search engine to find a good restaurant, the system may provide personalized assistance such as "I think you want to find a restaurant. Do you want to use Yelp? It has reliable reviews".

We describe data collection in Section 5.3.2 and observations from the collected data in Section 5.3.3. Intention recognition is discussed in Section 5.3.4. We describe the potential personalized interaction in Section 5.3.5.

5.3.2 Data collection

We developed an Android Application to track a user's everyday application usage. We recruited Android smartphone users and investigated their application usage behavior. For each day, a user's activities such as what application is launched at what time was logged and uploaded (after the user's approval). Geographic information is also logged to help user recall what happened.

We collected smartphone logs from 6 participants – 3 Americans, 2 Indians and 1 Chinese. We interviewed participants by showing them groups of apps they were using at a certain time. They were asked to recall the high level goal for using those apps. The following is an example: we found a frequent occurrence for two apps – FITBIT and CHOMPSMS. The former is for counting how many steps the user walked everyday and the latter is for sending text messages. We were told that he competes with his mom in who walks more everyday; he checks Fitbit and sends text message to his mom about the result.

5.3.3 Observation and statistics

App-level difference
We collected a total of 118 logs from 6 participants. Each participant has their own application vocabulary (top 5 most frequent ones are listed in Tab. 5.6).

We investigated how the application vocabulary changes over time. We sorted log files by time and plotted application level type-token curve as shown in Fig. 5.12. Users 4 and 5 did not produce sufficient data.

Tab. 5.6: Top 5 most frequent applications.

Rank	User 1	User 2	User 3	User 4	User 5	User 6
1	Gmail	WhatsApp	Chompsms	WhatsApp	MMS	Facebook
2	Chrome	Contacts	Contacts	Contacts	Spotify	Chrome
3	Google Search	Phone	Google Search	MMS	Facebook	GTalk
4	GTalk	Skype	Game	Gmail	Chrome	Phone
5	Twitter	Chrome	Fitbit	Phone	Twitter	Gmail

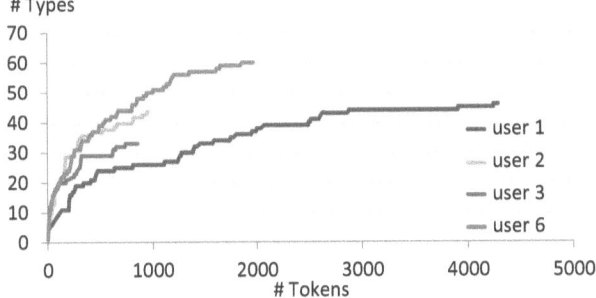

Fig. 5.12: Application-level type-token curve.

Episode-level difference

We introduce a definition of **episode** which is a group of applications serving a common high level goal. Within such a group, whether the order of the applications matters or not yields two different kinds of episode definition. For unordered episodes, for example, A wants to organize a dinner with B and C. A calls B and then sends a text message to C. This is the same as the scenario where A sends a text message to C first and then calls B. However, in the ordered version, these two scenarios are different. As we collect data, we do not have the high level goal of each application at hand. We use a time interval threshold to group applications – app X and app Y are grouped together if they happen within the threshold. Figure 5.13 shows the analysis of episode level type-token relationship. We removed episodes with only one application. In the figure, we show two users who have more logs than others. Solid lines correspond to ordered episode definition and dashed lines unordered. Obviously, the ordered version has more types of episodes than the unordered version. We focus on the unordered version of episodes. The most frequent episodes are shown in Tab. 5.7 (episodes with a single app are removed). As we can see, most of them are composed of the most frequently used apps in Tab. 5.6.

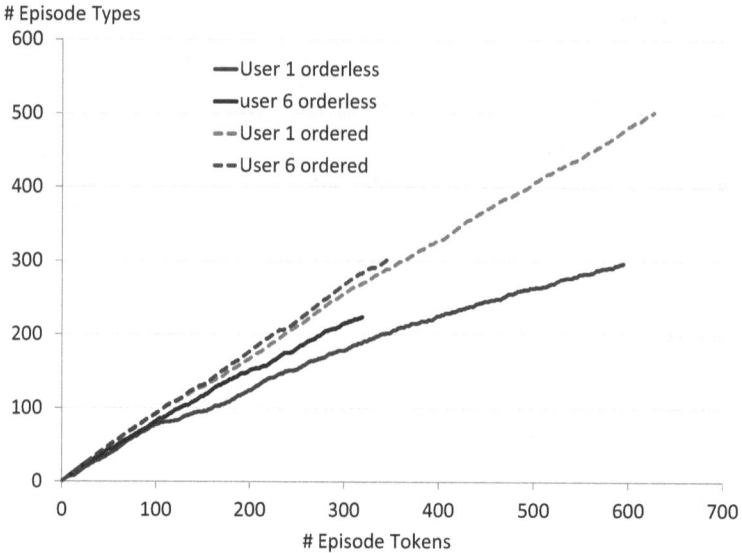

Fig. 5.13: Episodes-level type-token curve.

Tab. 5.7: Top 5 most frequent episodes (singletons removed).

Rank	User 1	User 6
1	Chrome + Google	Chrome + Facebook
2	Settings + GTalk	Phone + GVoice + PhoneBook
3	Chrome + Gmail + Google + GTalk + Twitter	Phone + PhoneBook
4	Gmail + Gtalk	Email + Facebook
5	Gmail + Google	Chrome + Facebook + Gmail

We also observe that, for the same episode, the intention of the user is different. For example, we observe many instances of "camera + text_message". By interviewing individual users, we understand that for user 1, he used this combination to tell his friend what he is doing; for user 2, he sent his friends pictures of academic talks he found on notice boards on campus.

5.3.4 Intention recognition

There are two different ways to recognize the intention of a user: explicit and implicit. For the explicit way, user may say "I want to schedule a dinner with friends". For the implicit way, the system needs to observe a few applications within a time interval to decide a list of possible intentions with probabilities. For example, a user may open "email" and "map" in a row. This may indicate that with 75 % chance this user wants to go to an event and does not know the exact location of the place; and 5 % chance this user is accidentally opening email and actually wants to open map. Implicit recognition may require maintaining a ranked list of possible hypotheses with probabilities and adjusting this list when more applications are launched. Communications within each application may help maintain the list. Research on dialog state tracking has led to a variety of proposed solutions (Williams et al. 2013; Henderson, Thomson & Williams 2014). The true label for each episode is difficult to obtain. We propose to work around this issue by using co-occurrence to recognize implicit intentions. For the explicit recognition, we assume that before launching any application, the system is already aware of what the user wants to achieve by receiving a complex verbal command from the user. This can be simulated by either (1) obtaining the user's label for each episode; or (2) defining episodes as "a set of applications grouped together".

5.3.5 Personalized interaction

Personalization has been studied in speech information systems (Partovi et al. 2008; Thompson, Goker & Langley 2004). Personalized recommendation/information is conveyed by the system in specific domains according to the user profile. Our work focuses on high-level communication of user intentions. This may help the dialog manager effectively structure a dialog flow using an appropriate sequence of subdialogs, each of which serves a specific goal in a given domain.

In this section, we discuss two potential use cases where a dialog system may benefit from knowing user intentions – (1) it can adjust its understanding models to favor language accordingly; (2) it can organize steps and suggest alternatives.

Model preparation

For different applications, people naturally have different ways of talking and different semantics to convey. Application-specific understanding models can handle the corresponding application, given sufficient training data. However, transition language among applications is hard to predict. Models such as application-level n-gram may be utilized to predict possible application in the next step. For example, if Chrome is always opened after Gmail, the system can expect the user to say "I want to send an email" at the end of using Chrome app. A different thought on building recognition models is this: as in episodes, we group/interpolate language models of those applications within the episode and run this interpolated model across all those apps without switching back-and-force among app-level models. The benefit of doing so includes: (1) reducing the effort(time, power) for switching understanding models; (2) solving the issue of what sentences to be included in app-level model training (e.g. will "I want to send an email" be in the interpolated model no matter which application it belongs to).

For each application, we build a skeleton grammar. We propose to sample the grammar to generate an artificial corpus for each individual application's language model estimation. For a given intention (e.g. a group of applications), language models for its corresponding applications are linearly interpolated either uniformly or with a weight trained on a development set.

We propose to evaluate the language we collected for each episode from participants by using the interpolated LM described above. We compare this with a baseline where interpolation is not performed. For the baseline system, the language model is trained on utterances cross applications.

Task organization and evaluation

There can be several paths to achieve a certain goal/intention. People have their own preferred ways among all the paths. This provides the opportunity to study (1) can a system provide smooth transitions among applications and (2) can a system influence users in terms of which path they choose. In this section, we discuss how people may be influenced if some alternative path outperforms their own. For example, a user sees many seminar talk posters on a board. He wants to tell his friends about some interesting ones. User A may take a picture and send the picture to friends. User B may send text messages about the details to friends. We may have time-efficiency as a global objective function to maximize. So perhaps user A is more efficient than user B. We want to investigate how to reveal such advantage to user B in natural language. There are a few ways to express the same meaning as shown below.

1. "You can use camera plus text message."
2. "I understand that you want to tell your friends about a talk. You can use camera plus text message."
3. "You can use camera plus text message. This will save you time."
4. "I understand that you want to tell your friends about a talk. You can use camera plus text message. This will save you time."

Compared to the first version, all others have additional information attached. For example, intention transparency can be conveyed through "I understand that you want to ..." and utility of the alternatives can be expressed by "This will save you ...".

Tailoring the information contained in a sentence for a dialog partner is observed in human-human communication and human-machine communication as well (Brennan 1996). For dialog systems, information presentation strategies have been studied (Walker et al. 2004). Verbal or non-verbal state transparency has been proved useful in human-machine communication (Pappu, Sun & Rudnicky 2013; Pappu et al. 2014).

5.3.6 Summary

We propose simulating cross-domain human-machine interactions by understanding how human use smartphone applications. Our preliminary results indicate that user behavior in terms of app-usage differs across different users, in terms of (1) the applications they use; (2) the tasks they want to perform; (3) the way they perform similar tasks. These observations imply that modeling the high-level intention of the user can improve the dialog system performance in terms of personalization and efficiency.

5.4 Conclusion

In this chapter we described an adaptation framework that learns user vocabulary, language and intentions. For lexicon adaptation, we found that expanding existing vocabulary is feasible and useful in reducing OOV rate and improving recognition accuracy. We designed a language adaptation framework to combine local adaptive ASR with cloud ASR to adapt to domain and users and maintain reasonable language coverage. We proposed to extend it to handle the situation where adaptation data has noise. We also proposed to investigate how to understand human intentions when they use voice applications to organize certain tasks. We simulate this process by analyzing smart device application usage.

Abbreviations

ASR   Automatic Speech Recognition
FSG   Finite State Grammar
LM   Language Model
OOL   Out of Language (word)
OOV   Out of Vocabulary (word)
POS   Part of Speech
SLM   Statistical Language Model

SWB   Switchboard (corpus)
WER   Word Error Rate
WSJ   Wall Street Journal (speech corpus)
xRT   Times Real Time (execution speed)

References

Anastasakos, T & Balakrishnan, SV 1998, 'The use of confidence measures in unsupervised adaptation of speech', *International Conference on Spoken Language Processing*.

Bacchiani, M & Roark, B 2003, 'Unsupervised language model adaptation', *Proc. of ICASSP*, pp. 224–227.

Baker, CF, Fillmore, CJ & Lowe, JB 1998, 'The Berkeley FrameNet Project', *Proc. of COLING*, Montreal QC, pp. 86–90.

Bazzi, I & Glass, J 2000, 'Modelling out-of-vocabulary words for robust speech recognition', *Proc. of ICSLP*, vol. 1, pp. 401–404.

Bellegarda, J 2004, 'Statistical language model adaptation: Review and perspectives', *Adaptation Methods for Speech Recognition*, vol. 42, pp. 93–108.

Bellegarda, J 2000, 'Method and apparatus for a speech recognition system language model that integrates a finite state grammar probability and an N-gram probability', Google Patents.

Bisani, M & Ney, H 2005, 'Open vocabulary speech recognition with flat hybrid models', *Proc. of Interspeech*, pp. 725–728.

Blaylock, N & Allen, J 2005, 'Generating artificial corpora for plan recognition', *International Conference on User Modeling* (UM).

Brennan, SE 1996, 'Conversations with and through Computers', *User Modeling and User-Adapted Interaction*, pp. 67–86.

CMU Pronouncing Dictionary 2015, (Wikipedia article). Available from: https://en.wikipedia.org/wiki/CMU_Pronouncing_Dictionary [17 January 2016].

Das, D 2012, *Semi-supervised and latent-variable models of natural language semantics*. PhD Thesis, Carnegie Mellon University.

Fiscus JG 1997, 'A post-processing system to yield reduced word error rates: Recogniser output voting error reduction (ROVER)', *Proc. of IEEE Workshop on Automatic Speech Recognition and Understanding*, Santa Barbara CA, pp. 347–352.

Gretter, R & Riccardi, G 2001, 'On-line learning of language models with word error probability distributions', *Proc. of ICASSP*, pp. 557–560.

Henderson, M, Thomson, B & Williams, J 2014, 'The second dialog state tracking challenge', *Proc. of SIGdial*, pp. 404–413.

Jelinek, F 1991, 'Up from trigrams! The struggle for improved language models', *Proc. of Eurospeech*, pp. 1037–1040.

Jelinek, F, Merialdo, B, Roukos, S & Strauss, M 1991, 'A dynamic language model for speech recognition', *Proc. DARPA Workshop on Speech and Natural Language*, pp. 293–295.

Kellner, A 1998, 'Initial language models for spoken dialogue systems', *Proc. of ICASSP*, pp. 185–188.

Klakow, D, Rose, G & Aubert, X 1999, 'OOV-detection in large vocabulary system using automatically defined word-fragments as fillers', *Proc. of Eurospeech*, pp. 49–52.

Knight, S, Gorrell, G, Rayner, M, Milward, D, Koeling, R, & Lewin, I 2001, 'Comparing grammar-based and robust approaches to speech understanding: A case study', *Proc. of Eurospeech*, pp. 1779–1782.

Lange ,P & Suendermann-Oeft, D 2014, 'Tuning Sphinx to outperform Google's speech recognition API.' *Proc. of the ESSV Conference on Electronic Speech Signal Processing*, Dresden, Germany, March 2014.

Leggetter, C & Woodland, P 1995, 'Maximum likelihood linear regression for speaker adaptation of continuous density hidden Markov models,' *Computer Speech and Language*, vol. 9, no. 2, pp. 171–185.

Lin, H, Bilmes, J, Vergyri, D & Kirchhoff, K 2007, 'OOV detection by joint word/phone lattice alignment', *Proc. of ASRU*, pp. 478–483.

Meteer, M & Rohlicek, J 1993, 'Statistical language modeling combining N-gram and context-free grammars', *Proc. of ICASSP*, pp. 37–40.

Milette, G & Stroud, A 2012, 'Professional Android sensor programming,' John Wiley & Sons.

Miller, G 1995, 'WordNet: A lexical database for English', *Communications of the ACM*, vol. 38, no. 11, pp. 39–41.

Mikolov, T, Chen, K, Corrado, G & Dean, J 2013a, 'Efficient estimation of word representations in vector space', *Proc. of ICLR Workshop*, arXiv:1301.3781.

Mikolov, T, Sutskever, I, Chen, K, Corrado, G & Dean, J 2013b, 'Distributed representations of words and phrases and their compositionality', *Proc. of NIPS*, pp. 3111–3119.

Mohri, M, Pereira, F & Riley, M 2002, 'Weighted finite-state transducers in speech recognition', *Computer Speech and Language*, pp. 69–88.

Morbini, F, Audhkhasi, K, Sagae, K, Artstein, R, Can, D, Georgiou, P, Narayanan, S, Leuski, A & Traum, D 2013, 'Which ASR should I choose for my dialogue system?' *Proc. SIGDIAL*, pp. 394–403.

Pappu, A 2014, *Knowledge discovery through spoken dialog*. PhD Thesis, Carnegie Mellon University.

Pappu, A, Sun, M, Sridharan, S & Rudnicky, AI 2014, 'Conversational strategies for robustly managing dialog in public spaces', *EACL Dialog in Motion Workshop*.

Pappu, A, Sun, M, Sridharan, S & Rudnicky, AI 2013, 'Situated multiparty interactions between humans and agents', *Proc. of HCII*, pp. 107–116.

Partovi, H, Brathwaite, RS, Davis, AM, McCue, MS, Porter, BW, Giannandrea, J, Walther, E, Accardi, A & Li, Z 2008, 'System for providing personalized content over a telephone interface to a user according to the corresponding personalization profile including the record of user actions or the record of user behavior', Patent, 2008.

Paul, DB & Baker JM 1992, 'The design for the Wall Street Journal-based CSR corpus', *Proc. of ICSLP*, pp. 357–362.

Pedregosa, F, Varoquaux, G, Gramfort, A, Michel, V, Thirion, B, Grisel, O, Blondel, M, Prettenhofer, P, Weiss, R, Dubourg, V, Vanderplas, J, Passos, A, Cournapeau, D, Brucher, M, Perrot, M, & Duchesnay, E 2011, 'Scikit-learn: Machine learning in Python', *Journal of Machine Learning Research*, vol. 12, pp. 2825–2830.

Placeway, P, Schwartz, R, Fung, P & Nguyen, L 1993, 'The estimation of powerful language models from small and large corpora', *Proc. of ICASSP*, pp. 33–36.

Prud'hommeaux, ET & Roark, B 2012, 'Graph-based alignment of narratives for automated neurological assessment', *BioNLP*, pp. 1–10.

Qin, L & Rudnicky, AI 2013a, 'Finding recurrent OOV words', *Proc. of Interspeech*, pp. 2242–2246.

Qin, L & Rudnicky, AI 2013b, 'Learning better lexical properties for recurrent OOV words', *Proc. of ASRU*, pp. 19–24.

Qin, L & Rudnicky, AI 2014, 'Building a vocabulary self-learning speech recognition system', *Proc. of Interspeech*, pp. 2862–2866.

Qin, L, Sun, M & Rudnicky, A 2011, 'OOV detection and recovery using hybrid models with different fragments', *Proc. of Interspeech*, pp. 1913–1916.

Qin, L, Sun, M & Rudnicky, A 2012, 'System combination for out-of-vocabulary word detection', *Proc. of ICASSP*, pp. 4817–4820.

Rosenfeld, R 1995, 'Optimizing lexical and N-gram coverage via judicious use of linguistic data', *Proc. of Eurospeech*, pp. 1763–1766.

Schaaf, T 2001 'Detection of OOV words using generalized word models and a semantic class language model', *Proc. of Eurospeech*, pp. 2581–2584.

Schalkwyk, J, Beeferman, D, Beaufays, F, Byrne, B, Chelba, C, Cohen, M, Kamvar, M & Strope, B 2010, 'Your word is my command: Google Search by voice: A case study,' *Advances in Speech Recognition*, pp. 61–90.

Sloboda, T & Waibel, A 1996, 'Dictionary learning for spontaneous speech', *Proc. of ICSLP*, pp. 2328–2331.

Sourceforge n.d. Available from: https://sourceforge.net/projects/cmusphinx/files/Acoustic%20and%20Language%20Models/ [17 January 2016].

Stolcke, A 2001, 'Error modeling and unsupervised language modeling', *NIST Large Vocabulary Conversational Speech Recognition Workshop*.

Strik, H & Cucchiarini, C 1999, 'Modeling pronunciation variation for ASR: A survey of the literature', *Speech Communication*, vol. 29, pp. 225–246.

Sun, H, Zhang, G, Zheng, F & Xu, M 2003, 'Using word confidence measure for OOV words detection in a spontaneous spoken dialog system', *Proc. of Eurospeech*, pp. 2713–2716.

Thompson, CA, Goker, MH & Langley, P 2004, 'A personalized system for conversational recommendations', *Journal of Artificial Intelligence Research*, pp. 393–428.

Twiefel, J, Baumann, T, Heinrich, S & Wermter, S 2014, 'Improving domain-independent cloud-based speech recognition with domain-dependent phonetic post-processing', in *Proc. of AAAI*, pp. 1529–1536.

Vertanen, K 2008, 'Combining open vocabulary recognition and word confusion networks', *Proc. of ICASSP*, pp. 4325–4328.

Walker, MA, Whittaker, SJ, Stent, A, Maloor, P, Moore, J, Johnston, M & Vasireddy, G 2004, 'Generation and evaluation of user tailored responses in multimodal dialogue', *Cognitive Science*, pp. 811–840.

Wang, Z, Schultz, T & Waibel, A 2003, 'Comparison of acoustic model adaptation techniques on non-native speech', *Proc. ICASSP*, pp. 540–543.

Wessel, F, Schluter, R, Macherey, K & Ney, H 2001,'Confidence measures for large vocabulary continuous speech recognition', *IEEE Transactions on Speech and Audio Processing*, vol. 9, pp. 288–298.

Williams, J, Raux, A, Ramachandran, D & Black, A 2013, 'The dialog state tracking challenge', *Proc. of SIGdial*, pp. 404–413.

Word2Vec n.d. Available from: http://code.google.com/p/word2vec/ [17 January 2016].

Xu, W & Rudnicky, A 2000, 'Language modeling for dialog system', *Proceedings of Interspeech*, Beijing, Paper B1-06.

Nava Shaked and Detlev Artelt

6 The use of multimodality in Avatars and Virtual Agents

Abstract: One area where multimodality is becoming essential is the design of Avatars and Virtual agents (A&VA). This multidisciplinary area combines several design capabilities that must work together. This chapter reviews the development of Avatars and Virtual Agents as part of the Human Machine Interaction (HMI) field, with an emphasis on the needs and challenges raised by user requirements and demands. Specifically, we are asking: How has Multimodality changed HMI to create a more versatile, personalized and accessible experience and reduced the gap between virtual assistance and live assistance?

Avatars and Virtual Agents, as we will show, create a "live"-like sensation and interaction experience thanks to the correct and smart fusion of the Multimodal capabilities. In fact, the more sophisticated the interface is and the more knowledgeable, the better the interaction between users and their Avatars. Virtual Agents are able to get the user to "task completion home run" if they can be personalized to the right extent and, no less important, if they are efficient.

We claim that it is possible to categorize the different existing multimodal interaction types for Avatars and Virtual Agents and propose a framework based on three types of "interaction relationship", with an example for each type.

6.1 What are A&VA – Definition and a short historical review

There is substantial evidence that avatars are not a phenomenon of the 20th century. Although the first "sophisticated" avatar appeared much later, the idea to make bodies as entertainment systems has existed for a very long time (Cassell 2001). The word "avatar" (origin in Sanskrit *avatāra* (Avatar 2015)) originally derives from Hinduism where it is a bodily manifestation of Immortal Beings or "the Supreme being" (Egen 2005). This definition, when applied to computers, means that an avatar is a representation in the virtual world and for some it can be seen as their "Incarnation" into the Internet (Egen 2005).

We will give a short overview of the most important milestones in the history of avatars and take a look at how they go hand-in-hand with specific technological developments.

We will also introduce Virtual Agents and their development in the digital media as representatives of service providers for self-service applications. To describe the difference between an Avatar and Virtual Agents we turn to Fox et al. (2015) who define avatars as virtual representations perceived to be controlled by humans and virtual agents as those perceived to be controlled by computer algorithms. Virtual representations of people in computer-mediated interactions can be categorized as avatars or agents.

"Avatars are distinguished from agents by the element of control: avatars are controlled by humans, whereas agents are controlled by computer algorithms. Hence, interaction with an avatar qualifies as computer-mediated communication (CMC) whereas interaction with an agent qualifies as human-computer interaction" (Fox et al. 2015, p. 5). Later in the chapter we will refer to this definition and suggest some additional criteria.

So what is so special about Avatars and Virtual Agents (A&VAs)? Humans by nature engage in complex representational activity involving speech and hand gesture, and they regulate that activity through social conversational protocols that include speech, eye gaze, facial expressions, head movement, and hand gesture. Casswell (2001) claims that, where social collaborative behavior is key, representing a system as a human is the correct interface. Her term for A&VAs is Embodied Conversational Agents (ECA). She argues that an ECA is an interface in which the system is represented as a person and in which information is conveyed to human users via multiple modalities such as voice and hand gestures, etc. Casswell (2001) was actually able to translate the rules of engagement into a table of conversational functions and their behavior realization which, in turn, could be used for ECA-based systems so that the user interface will be as similar as possible to human-to-human interaction.

We will examine all these interaction issues in relation to their supporting technologies but first let's look at the historical development.

6.1.1 First Avatars – Bodily interfaces and organic machines

The earliest attempts to model the body and bodily interfaces date back to the eighteenth century. Cassel (2001) claims that organicist automaton makers were driven by the question whether one could design a machine that could talk, write, interact in the way people do. Scientists wanted to know if machines could perform the same tasks as humans and to what degree they were able to perform them. This was the birth of artificial intelligence (AI) and the first steps towards the development of modern machines and computers.

The first machines were primarily developed for entertainment purposes, but soon they led to a completely new way of thinking. It was the French philosopher René Descartes (1596–1650), the "Father of Modern Philosophy", who in the seventeenth century first expressed the thought that animals and humans operate by mechanical principles. There is also some evidence that he believed that a body was the same as a machine (Mastin 2008).

One of the first organicist machines was invented by the Swiss-born watchmaker Pierre Jaquet-Droz in the 1770s. The "writing boy" is a life-sized doll that can move its arm to dip a quill in the inkpot and write text to 40 characters. The automaton comprises approximately 6000 parts and 40 replaceable interior cams that dictate the characters written (Hills 2013). Another automaton that achieved great popularity in Europe was a mechanical duck constructed by Vaucanson in 1738. It consisted of gold-plated copper and contained more than 1,000 parts including a digestive tract. Vaucanson's aim was to understand the bodily functions of human beings and find treatment options to heal diseases (Perkowitz 2004).

6.1.2 Modern use of avatars

The first digital avatars and agents appeared in the earliest days of computer development in the 1950s and 1960s. One of them was ELIZA, created by Stuart Weizenbaum. This computer program appeared in the early 1960s and emulated a Rogerian psychotherapist. ELIZA gave an illusion of intelligence and could answer simple questions (Weizenbaum 2000).

This program reminds us of what later became chat-boxes for customer service, which started to flourish in the 1990s as the Internet was released for public use. Chatterbots are still often integrated into the dialog systems of automated online assistants, with the ability to provide information to customers. During the 1980s and 1990s we also find voice-based Virtual Agents in call center IVRs for self-service – most popular in the banking, finance and public service domains. A well-known virtual agent figure was Julie from Amtrak. Over ten years ago Julie was assigned as Amtrak's friendly voice that you hear when you call their 1-800 number. It started a persona that was at first mostly associated with the voice enabled services and is now also used by Amtrak for Internet chatting and is associated with a woman's image that guides you through the company website ("Ask Julie", Amtrak 2016).

In the 1990s the first online avatars could be found as well. Early ones were simply animated characters mainly to attract more attention, while others presented information to video gamers, for example. In addition, the development of virtual and augmented reality technology – especially in gaming – required the user to choose an avatar to activate and, in most cases, play an extension of himself.

With the development of 3D graphics, gesture recognition and real-time image processing, the usage of virtual images representing the player is growing further. One of the most popular games using avatars is World of Warcraft. In this MMORPG (Massive Multiplayer Online Role Playing Game), the gamers choose a clearly combat focused Avatar. With about 7.1 million subscribers as of Q1 2015 (Statista 2015), World of Warcraft is currently the world's most-subscribed MMORPG. Digital avatars as virtual representations of real humans are also used in virtual worlds, such as Second Life.

What is common to all these examples is that the virtual image (avatar or agent) has human-like features and capabilities, with the ability to speak, move and understand what the user is communicating. A virtual avatar is able to conduct an interaction at some intellectual level (from light to sophisticated).

6.1.3 From virtual "me" to virtual "you"

Whereas the first organicist machine representations were of general human (or animal) figures and mostly not of a specific person, avatars in video games or virtual worlds represent a specific individual. They are used as a "virtual me", acting instead of the user in a virtual world. By the same token, we see in the last few years frequent usage of avatars as individual "virtual you" personalities, acting instead of customer service representatives.

What is common to all of them is that they can be "contacted" and interacted with in various ways using several communications channels as well as different input and output technologies, in conjunction with data transfer by apps or sensors – exactly as Caswell (2001) suggested. Some of these implementations require a high order of Artificial Intelligence technology to support them, such as NLP, machine learning, etc.

The ability of an avatar to generate face-to-face communication between real and virtual persons allows a much richer communication channel. It enables multimodal communication through both verbal and nonverbal channels such as gaze, gesture, spoken intonation and body posture (Nassiri, Powell & Moore 2004).

Furthermore, in a world in which mobility is a core topic, mobile devices, trackers, mobile sensors and smartphones are clearly connected to avatars. People create their own "virtual me" and "feed" it with multimodal data from sensors like fitness trackers or apps, so that they can measure weight loss, health or even sleep phases. There is no doubt that the interaction between humans and machines has reached a new level.

6.2 A relationship framework for Avatars and Virtual Agents

Our basic working assumption is that a Human-Machine relationship is taking place in the interaction and the aspiration is to achieve a dialog that enables task completion whatever the user task is.

In their work Rich, Sindner and Lesh (2000) discuss the theory of Collaboration, which is basically intended for Human-to-Human interactions, and apply it to Human-Machine interaction. "Collaboration" is a process in which two or more participants coordinate their actions towards achieving shared goals (Rich et al. 2000, p. 2). They introduce the concept of an agent independent of the application that communicates with the user in a three-way communication (agent, app, and user). The machine has been personified and "Collaboration" is now taking place between two users: one virtual and one human.

In our examination of theories of collaboration, we see that we can directly relate them to the concept of team work and we find three perspectives as described by Goldman and Degani (2012): the Humanistic View (teams with Human-only interaction), the Mechanistic view (in AI, for teams of Machines) and the Human-Machine view (joint teams).

If we accept the definition that Team Work means being responsive to each other's needs and mutually supportive to succeed in the joint plan, then we definitely get a sense of relationship that is being created between users (in our case between the Avatars and Virtual Agents and the human end user).

If this is the case, then let's examine and classify the types of relationships created in the dialog. We want this classification to be the basis for making a decision about which multimodal technologies or interaction types to choose for our Human-Machine Interface design.

We propose three main relationships types between human users and Avatars and Virtual Agents:
- Type 1: The Avatar as virtual me
- Type 2: Interaction with a personalized/specialized Avatar
- Type 3: Me and a random Avatar

6.2.1 Type 1 – The Avatar as virtual me

In this relationship type the Avatar is a mirror image of the user, a persona used to function as an extension for various purposes (playing a virtual game, personal assistance, medical avatar to be used in health applications, and a reflection of fitness and health application). The relationship is self-contained; personal activities are tracked without the need for external intervention. This type of Avatar is highly personalized according to the profile of the user and his preferences, with the ability to learn patterns and behaviors and the relationship can take on a high degree of intimacy. The preferred modalities will be inherent to the user and his profile and will not change very much.

In 1985 a video game called Ultima V was published. It was one of the first role-plays allowing the user to take on a different identity. The player starts as a regular character and tries to become an avatar. For the first time in gaming, an avatar appears. We already mentioned World of Warcraft (WoW), created in 2004 and today the world's most-subscribed MMOPRG. In the virtual universe of Azeroth every gamer can choose an avatar and personalize it, no matter whether he or she wants to act as a magician, elf or even monster (Ultima V 2015). But in these digital days, there are many more virtual worlds and a huge variety of avatars. Second Life, for example, developed by Linden Lab, is more than a game for its users. For them it's a kind of lifestyle or even an attitude towards life. Second Life users create avatars that can do anything that can be done in the real world: explore new things, chat, do business, create things. This virtual world is not a game as we know it – it's a parallel universe, just digital (Second Life 2015). It is very interesting how people act in virtual environments. Researchers have found that people tend to choose virtual characters that are a slightly "better" or an idealized version of themselves. People who have low self-esteem often choose avatars that like to socialize a lot. On the other hand researchers have also found, however, that users who create a more self-like avatar enjoyed the game more (Madigan 2015).

The other kind of avatars that belong to Type 1 are personal assistants that behave as an extension of the user for the purposes of self-monitoring, self-management and recording data related to a variety of activities. The popular applications of market leaders – such as Siri (Apple), Cortana (Microsoft), Ivee, Echo (Amazon) and Google Now (Google) – provide personalized services for mobile applications.

An avatar for elderly people called GeriJoy is a caregiving companion, built to address many of the unique challenges faced by seniors and their families. The GeriJoy Companion avatar is designed to be a supportive friend and caregiver. It is able to listen to – and remember! – what the user is saying, like names of grandchildren, favorite places, TV shows. It monitors the emotional states of the elderly such as feeling lonely or confused, and can provide engaging and supportive conversation.

Yet another example is Medical Avatar (Medical Avatar LLC 2016), an Internet site that provides online health management and maintenance services. It can identify health symptoms on the patient avatar while the user is interacting with a 3D image to create his mapping of his illness or complaints.

Of course both the GeriJoy and Medical Avatar services are supported with a backend of live agents and caregivers, but from an efficiency point of view, enabling self-service for at least some activities can save time by handling simple requests, especially for frequent users.

6.2.2 Type 2 – The interaction with a personalized/specialized avatar

This relationship type refers to the use of a specialized Virtual Agent that specializes in a certain type of activity such as: banking and financial services, HMO health agents, education, and government assistance. On the one hand these virtual agents represent a company, an organization or a service provider but, on the other hand, they are not random. They have prior knowledge of the user as their customer, sometimes on a routine basis. They know the profile of the user, his history and habits. They may be not as good as the personal avatar but in their specific line of business they can make predictions and next best offers. They can also connect generally available data with personal data to maximize service.

Nina is an intelligent multichannel customer service virtual assistant – a platform able to connect to IVRs, Internet sites, chat boxes and mobile apps. It can take on different personas depending on the enterprise it represents. One example is Ines, Nespresso's (AI4US 2016a) virtual agent. She has prior knowledge about the user, his account and his regular purchases. In addition, the user can ask her anything about the products and services offered. She gives users access to their information, helps users register with The Nespresso Club, and assists with connection issues and access to the Nespresso site.

The advantage of a personalized, known agent across all channels of the enterprise is clear. It is meaningful both in light of the theory of collaboration mentioned earlier as well as the notion of familiarity and personalization of services, which is crucial for the customer experience as a whole.

Online Banking is growing and is directly connected to a multichannel approach whereby users can access their account information and perform actions from various digital channels such as Internet, mobile app, telephone, chat, etc. Having an agent that is common to all channels and familiar with the customer's profile and habits is an advantage that is leveraged to create a holistic, seamless customer experience across every encounter.

To clarify, the difference between Type 1 and 2 in our definition is that Type 1 Avatars are an actual extension of the user and his daily habits be it in his gaming, his virtual world or his own personal assistants on his devices, where his profile resides and he can select his own preferences. In Type 2 we are talking about a virtual agent who is not under the control of the users but rather operated by a service organization, yet has prior knowledge of and can make personalized offers to the user.

6.2.3 Type 3 – Me and a virtual agent that is random

The third type of relationship is a relationship between the user and a Virtual Agent with whom he is not familiar, a random interaction with a virtual representative who contacts the user while he surfs the Internet or through an online segmentation and targeted advertisement campaign.

Here the user is mostly in a passive and less cooperative mode. For design purposes these agents must be very interesting, articulate and very engaging to get user attention and response.

Essentially these random virtual agents can be defined as virtual personas answering real-time questions asked on a website, at click speed, without pauses, 24 hours a day, completely automatically. These VAs can engage in a dialog and help customers make a decision.

The first Avatars used in marketing were only able to give information, with no ability to engage in two-way interaction. The Microsoft assistant Clippy (Cheezburger Inc. 2016) is a good example. When the user started typing, the assistant with googly eyes popped up and offered help. But things have changed a lot since then. The digital revolution and the availability of various interaction technologies have enabled Avatars and users to interact with each other more naturally and effectively. People seem to like interacting with Virtual Agents, as they provide a social dimension to the interaction. Humans willingly ascribe social awareness to computers (Nass, Steuer & Tauber 1994), and thus interaction with Virtual Agents follows social conventions, similar to human-to-human interactions. This social interaction both raises the believability and perceived trustworthiness of the agents as well as increases the user's engagement with the system. Another effect of the social aspect of agents is that presentations given by a Virtual Agent are perceived as more entertaining and more agreeable than the same presentations given without an agent (Van Mulken, André & Müller 1998).

Virtual Agents of this kind are also called "Chatbots". There are many good examples of Chatbots being used in marketing or customer service that offer more than just information. A current example is ALEXA – the virtual assistant from Amazon. Integrated into Amazon ECHO for the first time, Amazon plans to release the assistant for tablets and smartphones as well. It works via Amazon web services and requires a good Internet connection. It offers weather and news from a variety of sources and will play music from the owner's Amazon Music accounts. ECHO will respond to your questions about items in your Google calendar and other interfaces. Like Siri or Cortana, Alexa can be controlled by speech. Nevertheless ECHO is not a human-like figure; it is a cylinder-shaped object to be used in the home or office.

The site www.chatbots.org lists a wide range of customer service Virtual Agents – like Ines based on the Nina platform (AI4US 2016b) – with different gender persona and visualization. Some examples are Agent Striker for IGN Entertainment, Nathan for Symantec, Charlie for AT&T, Mr. Bibendum for Michelin, Cloe for VirtyOz, and many more.

In conclusion, defining these three framework types in this section is important in order to characterize the different types of Avatars and Virtual Agents and distinguishing them by means of the relationship they form with the user. These distinctions will also be important later when evaluating the quality of a given platform for creating the evaluation matrix in Section 6.3.3

6.3 Multimodal features of A&VA – categorizing the need, the challenge, the solutions

Up to this point, we have described the evolution of Avatars and Virtual Agents and suggested a relationship framework to describe the different types of human-machine relations. Next we will discuss the usage of Multimodal interaction in Virtual Avatars and Agents and suggest a methodology to facilitate the design and fusion of technologies for different A&VA applications.

The context for this discussion is very straightforward. Human-machine interaction has been changing to accommodate to the new digital era, with Multimodal interaction technologies helping to successfully bridge the gaps. Our user is living in a mobile environment based on state-of-the-art digital capabilities and is looking to interact intensively in order to get information, collect data, get customer service and perform actions.

6.3.1 About multimodal interaction technologies

Multimodal systems combine two or more input and output modes for human-machine interaction – such as speech, handwriting, gesture and touch, sketch, eye movement, facial expression, and so on. "This new class of interfaces aims to recognize naturally occurring forms of human language and behavior, which incorporate at least one recognition based technology (e.g., speech, pen, vision)" (Oviatt & Designs 2012, p. 1)

Fortunately today system designers are able to successfully offer many of these modes relying mainly on the mobile infrastructure and the availability of communication networks. Nevertheless, as we will show, the offering of interaction modalities is directly dependent on the functionality and the nature of the application.

Oviatt and Cohen (2015) have created a table containing interaction modes according to different mobile platforms and infrastructures. Some are already popular while others are still in the early adopters stage. But it is clear that with the IoT (internet of things) revolution and 5G cellular networks, the incorporation of more and more modalities will be a requirement of the users themselves. This table (Oviatt & Cohen 2015, p. 136–137) presents applications that use Multimodal interaction. It shows the status of actual combinations and implementations that already exist in the market. For each of these combinations there is an industry-line as well as actual product providers. One of the industry lines is Virtual Assistance, which is a sub-area of the A&VA field. A careful look reveals that Virtual Assistance includes the largest number of possible modalities to be used: Voice, Sketch, Touch, Gesture, Handwriting, Mouse, Gaze, Face tracking, Keyboard, Buttons, Head position, Torso position, Eye tracking, Arm position, Face recognition, Fingerprint, Iris, and Voice biometrics (Oviatt & Cohen 2015, p. 136–137).

In fact we claim that in an Avatars & Virtual Agents platform we will be using the maximal number of modalities – higher than in any other industry line. This highly multimodal approach addresses the goal of creating more flexible, efficient, challenging and easy-to-use interactions for different user populations and the need to accommodate the different types of A&VA relations described in Section 6.2 above.

6.3.2 Why use multimodality with Avatars?

Whereas the first Avatars were mostly attention-grabbing "personas" limited primarily to providing information, modern Avatars offer much more. Whether in gaming, education, personal assistants or even in healthcare, customer service and marketing, the user experience and the task completion rate of the interaction is crucial to the success of the platform. To demonstrate the importance of this issue, let's consider the following examples:

A user is logged-in, playing a video game using his avatar. The interaction between him and his avatar takes place using keyboard (for chat) and joystick and perhaps he is using video and voice too. If the interaction fails because, for example, his camera or keyboard drops out, an error will occur. In some cases he will be logged out and his session will end, which is disappointing but not critical because he can log in again.

But if an interaction is taking place between a user and his avatar on a healthcare site (such as a Doctor's site: www.MedicalAvatar.com) and in the middle of a personal checkup the application halts, the user might feel very uncomfortable regarding his health information. In another example, if you are calling a service center requesting information and the virtual agent is unable to understand you, keeps getting into an error loop because the speech engine is failing and the interaction is cumbersome – there will be no resolution of the request and the user experience will be disappointing. It has been established that multimodal ways of interaction are, most of the time, easier to use, require less training, are robust and flexible, as well as faster, more efficient, and, of course, supports new functionality (Dumas, Lalanne, Oviatt 2009).

So how can we design a satisfactory interface for A&VA using multimodal capabilities? From an interaction point of view we claim that three factors must be analyzed and modeled before designing a solution.

6.3.2.1 The user

Identifying the target user is a key factor. There's a difference whether a child or an adult is interacting with an avatar, or if the user is elderly or an injured or disabled person. Each of the user types mentioned will prefer one or more ways of interaction. Children will mostly communicate by speech, text, touch and gesture, but an elderly person may use eye position, voice or gesture. Other issues concern the output technologies preferred by each of the multimodal interaction user types: textual or image output, voice or cues, as well as how the data is presented.

6.3.2.2 The purpose

In his discussion Traum (2008) establishes the relationship between the behavior of people during a human-to-human dialog and a person's interaction with "virtual humans" (the term he coined for Avatars and Virtual Agents). Traum claims that the construction of dialogs with virtual humans can be based on similar cornerstones to human-to-human dialog, quoting Allwood's (1995) social activities parameters:

- Procedures: type, purpose, function
- Roles: competence, obligations, rights
- Instruments: machines, media
- Other physical environment parameters

But Traum also claims that we must consider that the behavior of people might be affected by the fact that they are conversing with a virtual character and this fact needs to be taken into account in the design of the dialog: "Purely human data can be used both as a starting point for analysis and implementation of virtual humans, and also as an ultimate performance goal, however it may not be the most direct data, especially for machine learning." (p. 300).

The next criteria concerns the task the user is trying to complete using the Avatar: Is the Avatar being used to get information, for Q&A, for education, diagnostics or to provide data to someone else. We can differentiate among three types of use cases: active, passive or mixed use.

Passive use. The user only receives passive information such as, for example, in transportation when an avatar is giving schedule times for buses or trains. This Avatar is usually a persona representing the company on its Internet site.

Active use. The user actively provides data to the Avatar or Virtual Agent. For example, using a fitness app with an avatar of "myself", giving data about food habits, about training, about running distance or sleep behavior by using some sort of fitness tracker or other sensors. The user initiates the action and actively engages with the virtual persona.

Mixed use. The user is not only providing data but also receiving information from the Avatar or Virtual Agents. Diagnosis solutions are a good example of mixed usage. The patient provides data to the Avatar using speech, touch, keyboard or sensors and gets information from it, e.g. a diagnosis or medical advice. If data is provided to the Avatar or Virtual Agent, usually two or more ways of interaction are used. The more extensively someone is giving data and information, more ways of interaction are used.

6.3.2.3 The environment

The interaction modality is also influenced by the usage environment. Users do not usually use speech in a very noisy environment or eye tracking when it's dark outside. Environment is not only the surrounding and ambience. It also includes the platform on which the virtual agent is used. Mobile devices allow for easier interaction in some respects while wearables or gaming devices have other restrictions and considerations. In fact, as claimed by Lawo, Logan and Pasher (2016, to appear), the notion of wearable ecology describes the environment in which we put the user and the technology in an intimate relationship. In their article they discuss this environment as empowering the user on the one hand, but as limiting and restricting on the other.

To sum up, in the first part of the section we claimed that A&VA will use as many modalities as possible. In the second part we clarify that the three criteria of User, Purpose and Environment must be factored in when selecting the appropriate modalities for an optimal multimodal interface design.

6.3.3 Evaluation of the quality of Avatars and Virtual Agents

This section addresses the research question of how to determine the optimal interaction design for A&VA. Let's review our argument so far; In Section 6.2, we examined a set of classification methods to characterize the user-machine relationship and established a relationship framework based on three types: Type 1: Avatar as virtual me; Type 2: Interaction with a personalized/specialized Avatar; and Type 3: Interaction with a random Avatar. In Section 6.3.2 we established the criteria to determine which modalities will fit different application and use cases depending on User Type, the Purpose of the application, and the Usage Environment.

The aim of this section is to present the set of features for evaluation. This set of features will help us best assess the overall quality of the Avatar, taking into consideration all the factors we have discussed above – and some others, which are related to performance measurement.

We suggest an evaluation framework which is based on a set of assessment features and directly linked to the three user-Avatar relationship types. For each relationship type the priority and weight of the feature is scaled and graded as L (low), M (medium), or H (high). This matrix of features, relationship type and grading, provides a novel approach to the evaluation of an avatar or virtual agent (see Tab. 6.1). For each feature we ask how crucial it is for the Interaction Type. Some of the features are based on general quality assurance best practices while others are specific to the A&VA platform.

The following is a brief description of the matrix basics, focusing only on those features that are central to the chapter's arguments. **Clearly, the matrix needs to be further developed and detailed.**

- **Ease of Use:** This feature is concerned with how friendly and easy to navigate the interface is, and if it requires a sophisticated learning effort. We claim that for all types this feature is crucial, and, therefore, is graded as High.
- **Visualization:** This feature deals with the external look of the A&VA, the app – whether there is a human-like agent, an image, or just an app user interface. How likeable is it and how does it represent the user or the enter-

Tab. 6.1: The evaluation matrix.

Feature	Sub-feature	Type 1	Type 2	Type 3
Ease of use: Task completion		H	H	H
Visualization: The look of the figure/app	General	H	H	H
	Of the Personas	H	M	L
Privacy features		H	H	M
Language support		L	L	H
Tasks available		Specific tasks	Mixed tasks	General tasks
Personalization		H	H	L
Error recovery capability		M	H	M
Low latency		H	H	H
Multimodal technology usage	Multiple I/O technologies	H	H	M
	Usage of sensors	H	H	L
Other technology support	Gamification	H	H	H
	AI/Machine learning	H	H	L
	Big Data analytics	H	H	L

prise? We claim that this feature is, in general, highly important to all types of A&VAs.

- When the A&VA is a **Persona** (human-like image), how important is it to the interaction and to the engagement of the user to create a remarkable user experience? Well, for Type 1, which is the extension of the user in an avatar ("virtual me"), it is of very high priority. For the personalized Virtual Agents it has medium importance since, after all, it is a reflection of the enterprise it is representing. For the random type (3) the visualization is of low value as long as it is agreeable and easy to use.
- **Privacy of information** is a grand and complex issue that cannot be covered here. We will simply say that in, our opinion, it is not of the same importance to all Types. In case of a random agent which does not possess any prior information, it will be of lesser importance while for Types 1 and 2 it must be a high priority issue.
- **Language Support** is a feature that enables the A&VA to support many languages and accommodate multi-language speakers. This is highly important

for the Type 3 random agent, which needs to support several target audiences without prior knowledge of their preferred language. This will not be necessary for Types 1 and 2, where the basic profile and preferences of the user are known in advance.

- **Diversity of available tasks** relates to how important it is that the A&VA will be able to offer an array of services of the following types: specific personalized ones, general tasks, or a mixture of both. How much emphasis should be placed on learning the user in real time and providing services based on his habits and behavior? We claim that for Type 1, personalized tasks are critically important, while for Type 2 a mixture of personalized and general tasks has to be supported. Type 3 requires only general tasks.

- **Error recovery** is the ability of the A&VA to gracefully recover from a dialog mistake or misunderstanding rather than create an infinite loop and a bad interaction experience. Error recovery is not easy and requires NLP as well as data analytics technologies, but it is highly important especially in a dialog with a virtual agent representing an enterprise in a commercial interaction. Failure to create a smooth interaction – especially if the agent has prior information regarding the user – is a problem. Another feature important to user experience is **Latency**, i.e., the delay between input into the system and the appearance of the desired output. Latency greatly affects the quality of the interaction as well as the flow of the dialog. As a result, low latency (the reaction in less than a few seconds per interaction) is graded as High for all three types.

- We have already established the importance of **Multimodal Interaction Technologies** in A&VA design, claiming that A&VAs can incorporate the largest number of input/output technologies into the user-machine dialog, whether that dialog be passive or active (see Section 6.3.1). We claim that for Types 1 and 2 multimodal interaction technologies are crucial, while for Type 3 it is of Medium priority – good to have but not critical since the expectation for interaction from a random agent is not as high as from a personalized agent or avatar. When we drill down to the usage of sensors for input information and for output feedback, once again this is of high priority to Types 1 and 2 and less so for the random Type 3.

- Other supporting technologies also determine the quality of the interaction and the level of the user experience. **Gamification,** i.e., adding gaming components and rewards for task completion and improvement into the A&VA-user dialog, is highly recommended for Type 1 Avatars and recommended for personalized Type 2 agents. For Type 3 it can be used effectively to engage the user with a new offering.

- As mentioned above, **Data Collection and Analytics** also play an important role in the process of personalization. Analytics facilitate the recognition of user behavior patterns and the prediction of future user behavior. It is highly critical for Types 1 and 2 and much less for Type 3.
- **Machine learning algorithms and AI capabilities** are necessary for an A&VA to interact in a natural, human-like way, leveraging the user experience and generating cooperation from the side of the user. To achieve this, the interaction must be monitored continuously over time, with the ability to collect, store and analyze the data and then run **Machine learning AI algorithms** to process the interaction and learn to improve it.

The matrix described above gives us a glimpse into the quality considerations relevant to A&VAs. It also includes some suggestions for a new approach to grading and scaling, but the subject definitely requires further elaboration and research.

6.4 Conclusion and future directions: The vision of A&VA multimodality in the digital era

How is HMI influenced by the new technological developments of the current Digital Era? This question is essential to the understanding of the future of the interaction technologies that form the basis for the abilities of A&VAs to communicate in a human-like manner and incorporate AI.

The very existence of Avatars and Virtual Agents depends on a successful user interface and is closely related to the whole theory of human-machine interaction. However, we will not review here the timeline of human-machine interaction development, but will jump straight to the current period and discuss the implications of the Digital Era.

Jill Shepherd (2004) claims that the digital era is characterized by technology that increases the speed and breadth of knowledge turnover within the economy and society. "The Digital Era can be seen as the development of an evolutionary system in which knowledge turnover is not only very high, but also increasingly out of the control of humans, making it a time in which our lives become more difficult to manage." Furthermore she claims that "the Digital Era has changed the way we live and work by creating a society and economy that is geared to knowledge, whether that knowledge is content-laden and therefore scientifically factual, or instead is content-free and therefore reliant on emotions, or indeed any combination in between." (Shepherd 2004, p. 2)

In this era people belong to active social and economic communities and the effect on all members of these communities is their aspiration to gain as much knowledge as possible. As technological functionality becomes more knowledge-based, we become more dependent on knowledge retrieval technologies and on-line devices. Thus understanding the Digital Era in terms of evolution means enrolling technologies to provide service and knowledge in every area and using technology to build assistive Avatars or Agents is an excellent subject of this revolution.

Meisel (2013) suggests that personal assistance apps like Siri point to the growing sophistication of the human relationship with digital systems. He claims that "user interfaces will continue to improve over time, driven by improvements in the underlying technology, faster digital processors, improved connectivity with computer networks and user feedback." (Meisel 2013, p. 27).

On the one hand this change has led to new ways of interaction between humans and Virtual Agents and Avatars, and, on the other hand, to new interaction types and interfaces affected by the following four major leading trends that are highly relevant to our topic:

Infrastructure Technologies. The data availability revolution is powered by big data, cloud storage, hosted services and innovative user interface technologies. These developments have made it possible to go to the next level in terms of dialog systems and self-service. From information-giving avatars in the early 1980s to today's mobile personal assistants that support user input and interaction of two or more modalities (Oviatt & Cohen 2015).

Mobile Technologies. Mobility as a whole and the smartphone in particular that is found in virtually every pocket requires the development of supporting interfaces to fit the needs of all users. Mobile manufacturers provide multimodal interfaces and Avatar applications as basic offerings to the users. We see the fusion of data, sensors and input/output technologies into mobile devices in order to optimize their performance and the quality of interaction that they provide. For example, using the GPS sensor to locate the user, to count steps and then transfer this data to the Avatar can create a data routine that alerts the user in a fitness application. 3G and 4G capabilities have enabled higher and higher data rates over the last ten years. Streaming got more powerful and the usage of apps, chat and other smart functions predominate telephone functions (Rainie & Poushter 2014). Behold! The second digital revolution is coming! 5G networks will create an even stronger effect – making every device a communication channel and allowing IoT applications. Mobile devices have become the preferred multimodal platform for interacting with Virtual Agents and Avatars by enabling the usage of virtual communications through apps, sensors and mobile solutions.

Socio-economic phenomena. From Babyboomers (born 1946–1965), to Generation X (born 1965–1980) to Generation Y, so called Millenials (born after 1980), "smart" technology – and especially mobile devices – is integrated into their work and private lives. Eight out of ten young people aged 18–30 use a Smartphone, 76 % own a Notebook and every third young adult possesses a tablet (Heuzeroth 2014). The digital revolution has taken place and technology, apps, mobile devices and avatars are our daily companions. As the survey "Generation #Hashtag" by Bain & Company shows, as many as 2/3 of German users prefer digital media and use mobile devices (Kunstmann-Seik 2015). The social changes brought about by the growth of these technological changes are reflected in our habits and behaviors, as well as in the way we communicate, consume products, socialize and interact. That means that communicating with your PA, your bank's virtual agent, or your health avatar should be an easy task, with a friendly user interface that fits anywhere, anytime.

The elderly population, which is growing massively, is creating a new balance in market forces on the interaction domain, requiring the adaptation and adoption of new care methodologies and technologies. The same is also true with children. The growing use of A&VAs in care-taking and assistive, as well as educational, applications is driving the adoption of multimodal interfaces.

In conclusion, the technological developments that are an essential part of this era have cultivated and essentially forced a change in the way Humans and Machines interact and thus has brought new interface technologies to human-machine interaction in general, and to Avatars and Virtual Agents specifically.

Abbreviations

A&VA	Avatars and Virtual agents
HMI	Human Machine Interaction
AI	artificial intelligence
MMORPG	Massively Multiplayer Online Role Playing Game
NLP	Natural Language Processing
IoT	Internet of Things
WoW	World of Warcraft
HMO	Health Maintenance Organization
MMI	Multimodal Interaction
I/O	Input output
ECA	Embodied Conversational Agents

References

AI4US 2016a, 'Ines (Nespresso Club)', *Chattbots.org*. Available from: https://www.chatbots. org/virtual_agent/ines1 [17 January 2016].

AI4US 2016b, 'Chatbots by nuance', *Chattbots.org*. Available from: https://www.chatbots.org/ developer/nuance/ [17 January 2016].

Allwood, J 1995, *An activity based approach to pragmatics*, Technical Report (GPTL) 75, Gothenburg Papers in Theoretical Linguistics, University of Göteborg.

Amtrak 2016, *Ask Julie*. Available from: http://www.amtrak.com/about-julie-amtrak-virtual-travel-assistant [17 January 2016].

Avatar 2016, (Wikipedia article). Available from: https://en.wikipedia.org/wiki/Avatar [7 October 2015].

Cassell, J 2001, 'Embodied conversational agents: representation and intelligence in user interfaces', *AI magazine*, vol. 22, no. 4, pp. 67.

Cheezburger Inc. 2016, 'Clippy', *Know your meme*. Available from: http://knowyourmeme.com/ memes/clippy [17 January 2016].

Egen, S 2005, *The history of avatars*, iMedia Connection.

Fox, J, Ahn, SJ, Janssen, JH, Yeykelis, L, Segovia, KY, & Bailenson, JN 2015, 'Avatars versus agents: A meta-analysis quantifying the effects of agency on social influence', *Human-Computer Interaction*, vol. 30, issue 5, pp. 401–432. Available from: http://www.tandfonline.com/doi/abs/10.1080/07370024.2014.921494 [25 October 2015]

GeriJoy 2015, *Senior living*. Available from: http://www.gerijoy.com/senior-living.html [29 July 2015].

Heuzeroth, T 2014, *Generation Y fühlt sich von digitaler Welt gestresst*. Available from: http://www.welt.de/wirtschaft/article134497791/Generation-Y-fuehlt-sich-von-digitaler-Welt-gestresst.html [27 July 2015].

Hills, S 2013, *Was this automaton the world's first computer? Incredible mechanical boy built 240 years ago who could actually write*. Available from: http://www.dailymail.co.uk/news/article-2488165/The-worlds-Mechanical-boy-built-240-years-ago-engineered-act-writing.html [5 August 2015].

Kim, J 2013, *Health Buddy: Fitness-Avatar wird fett bei zu wenig Bewegung*. Available from: http://de.engadget.com/2013/08/20/s-health-buddy-fitness-avatar-wird-fett-bei-zu-wenig-bewegung/ [29 July 2015].

Kunstmann-Seik, L 2015, *Bain-Studie zur digitalen Mediennutzung: „Generation #Hashtag" setzt auf neue Medienformate*. Available from: http://www.bain.de/press/press-archive/generation-hashtag-setzt-auf-neue-medienformate.aspx [1 July 2015].

Lawo, M, Logan, R, & Pasher, E 2016, 'Wearable computing – A media ecology approach and the context challenge' to appear in *The Design of Mobile Multimodal Interfaces*, eds N Shaked & U Winter, DeGuyter, NY.

Lloyd, D 2013, *Mobile Virtual Agents for Self-Service Banking*. Available from: https://www.bai.org/bankingstrategies/article.aspx?Id=2f247148-b5e3-431d-92a9-d3406e253c73 [29 July 2015].

Madigan, J 2015, *The psychology of video games*. Available from: http://www.psychologyofgames.com/author/jamie-madigan/ [29 July 2015].

Mastin, L 2008, *René Deacartes*. Available from: The Basics of Philosophy http://www.philosophybasics.com/philosophers_descartes.html [17 January 2016].

Medical Avatar LLC 2016. Available from: https://www.MedicalAvatar.com [17 January 2016].

Nass, C, Steuer, J & Tauber, ER 1994, 'Computers are social actors', *Proceedings of the SIGCHI Conference on Human Factors in Computing Systems*, ACM, pp. 72.

MedicalAvatar 2015, Available from: http://www.medicalavatar.com [17 October 2015].

Nassiri, N, Powell, N & Moore, D 2004, 'Avatar gender and personal space invasion anxiety level in desktop collaborative virtual environments', *Virtual Reality*, vol. 8, no. 2, pp. 107–117.

New Hope Media LLC 2013, *To-Do List Apps for ADHD Kids and Adults*. Available from: http://www.additudemag.com/adhd/article/8698.html [29 July 2015].

Nuance 2015, *Self service with an intelligent touch of humanity*. Available from: http://www.nuance.com/for-business/customer-service-solutions/nina/index.htm [29 July 2015].

Oviatt, S & Cohen, PR 2015, *The paradigm shift to multimodality in contemporary computer interfaces*, Morgan & Claypool Publishers.

Oviatt, S & Designs, I 2012, 'Multimodal Interfaces,' in *Handbook of Human-Computer Interaction*, ed J Jacko, (3rd ed), Lawrence Erlbaum, New Jersey, pp. 405–430.

Perkowitz, S 2004, *Digital People: From Bionic Humans to Androids*, Joseph Henry Press.

Piccolo Picco Ltd 2015, *Emerging nations catching up to U.S. on technology adoption, especially mobile and social media use*. Available from: https://itunes.apple.com/us/app/fitness-avatar-exercise-trainer/id942101272?mt=8 [29 July 2015].

Rainie, L & Poushter, J 2014, *Emerging nations catching up to U.S. on technology adoption, especially mobile and social media use*. Available from: http://www.pewresearch.org/fact-tank/2014/02/13/emerging-nations-catching-up-to-u-s-on-technology-adoption-especially-mobile-and-social-media-use/ [27 July 2015].

Second Life 2015 (Wikipedia article). Available from: https://de.wikipedia.org/wiki/Second_Life [29 July 2015].

Shepherd, J 2004, 'What is the digital era?', in *Social and economic transformation in the digital era*, eds G Doukidis, N Mylonopoulos N Pouloudi, pp. 1–18.

statista 2015, *Number of World of Warcraft subscribers Q1 2005–Q1 2015*. Available from: http://www.statista.com/statistics/276601/number-of-world-of-warcraft-subscribers-by-quarter/ [27 July 2015].

Traum, D 2008, 'Talking to virtual humans: Dialogue models and methodologies for embodied conversational agents', in *Modeling communication with robots and virtual humans*, eds I Wachsmuth and G Knoblich, LNAI 4930, Springer-Verlag, Berlin, Heidelberg, pp. 296–309.

Ultima V: Warriors of Destiny 2015 (Wikipedia article). Available from: https://en.wikipedia.org/wiki/Ultima_V:_Warriors_of_Destiny [29 July 2015].

Van Arsdale, J 2014, *Fitness hero presents – 'Avatars'*. Available from: https://www.indiegogo.com/projects/fitness-hero-presents-avatars#/story [29 July 2015].

Van Mulken, S, André, E & Müller, J 1998, 'The persona effect: how substantial is it', *People and Computers XIII: Proceedings of HCI*, vol. 98, pp. 53–66.

Weizenbaum, J 2000, *Die Macht der Computer und die Ohnmacht der Vernunft*, 11th edn, Suhrkamp Verlag GmbH, Frankfurt.

Brion van Over, Elizabeth Molina-Markham, Sunny Lie, and
Donal Carbaugh

7 Managing interaction with an in-car infotainment system

Abstract: This work investigates trouble in multimodal turn exchange between an
in-car infotainment system and human interactants. The trouble is linked to a lack
of crystallization of norms surrounding the turn status of non-speech sounds as
well as misalignment on culturally constituted and variable indicators of upcoming transition relevance places. Four interactional adaptations employed by users
in order to accomplish their goals for the interaction despite the trouble are identified, as well as norms governing user interaction with the system, and cultural
premises that inform that interaction. The work concludes with a discussion of
considerations for future multimodal system design.

7.1 Introduction

Imagine a conversation wherein you ask a friend if they would like to see a movie.
The friend replies "what concert did you want to see?" Your first thought might be,
"What? How is that relevant to the question I just posed? Why might they have said
that? Did they mishear me? Perhaps they were implying they would rather go to a
concert than a movie?" Humans routinely use this kind of answer, one that flouts
Grice's (1975) maxim of relevance to accomplish communicative goals like conversational implicature. However, what are we to make of the following interaction
between a human user and an in-car multimodal infotainment system?

Instance 1: Context FM Radio.

```
1   Participant: (Participant touches microphone button)
2   System:     (audible ding)
3               (0.6)
4   Participant: phone ca[ll
5   System:                [which station or channel do you want to
                hear?
```

Here we see much the same oddity as in the hypothetical conversation between
friends; the user asks to make a phone call and the system replies with a question
about what radio station they would like to hear. Except unlike our conversation
between friends, here, the user is not likely to wonder if the system would prefer

to listen to a radio station or might be attempting a conversational implicature. So how, as a user of a system like this, do we make sense of this interaction when the framework we use to interpret similar human speech no longer works? How did we get into this situation to begin with, and what do we do about it now that we are here?

Users of multimodal systems like this are often faced with these kinds of communicative challenges because machines are imperfect interactants and frequently fail to follow basic rules and principles for the governance of communication that human interactants generally follow with each other. This poses a problem for studying these human-machine interactions through the application of many existing theories of social interaction because of their reliance on the assumption of a model interactant who is competent in the culturally distinctive ways humans have developed for communicating with one-another. This model interactant follows particular rules for the organization and ongoing management of conversation that are based on the assumption that humans interact from a set of what some have suggested are universal principles (Sacks 1974; Sidnell 2001; Stivers et al. 2009). One prominent example of a theory that relies on the model interactant is Brown and Levinson's (1987) Politeness Theory.

Politeness Theory posits a Model Person (MP) that consists in a "willful fluent speaker of a natural language, further endowed with two special properties – rationality and face." (p. 58) This MP is "rational" to the extent that they have goals for their interaction, and a process through which the optimal means of achieving these goals are known and pursued. The MP has "face" to the extent that all speakers have wants, "roughly to be unimpeded and the want to be approved of in certain respects" (p. 58).

Brown and Levinson's MP is informed by Grice's (1975) model interactant who follows what he dubs the "cooperative principle." Interactants who obey the cooperative principle act in accordance with the following rule: "make your contribution such as it is required, at the stage at which it occurs, by the accepted purpose or direction of the talk exchange in which you are engaged." Grice goes on to specify four conversational maxims that he suggests govern conversation that occurs under the cooperative principle, including the maxim of quantity (give the most helpful amount of information), the maxim of quality (do not lie), the maxim of relation (make your contributions relevant), and the maxim of manner (make your contributions clear, brief and orderly). Of course, Grice does not claim that all speakers follow these rules. In fact, Grice suggests these maxims can be violated where the rule is simply ignored, or they can be flouted, where the rule is broken for a sought conversational effect.

The relevance of these theories to the present discussion consists in noticing that machines in interaction with humans routinely violate behaviors expected of the Model Person, or the interactant operating under the cooperative principle. This is because machines do not currently exhibit the kind of rationality presumed by Brown and Levinson, nor share the "face" concerns of human interactants. They may seek to operate within the bounds of the cooperative principle, if programmed to attempt to do so, but may violate maxims in obvious ways similar to how a human interactant might flout a maxim in order to accomplish a conversational implicature (Grice 1975), though the machine has no intention of doing so and users likely know that.

This means that some of our fundamental assumptions about social interaction become unreliable in human-machine interaction. And by extension, machines themselves may be found to be unreliable interactants for these very reasons. Two questions then become essential for us to pursue in order to better understand the dynamics at work in human-machine multimodal interactions. (1) How do humans interact with machines that are not assumed to be, nor able to operate as, fully culturally competent interlocutors? and (2) how do we manage moments when things inevitably go wrong in these interactions?

One area designers of such systems have often overlooked in their attempts to create ever more human-like interactants is the structure of turn exchange in conversation between humans and machines, with an eye toward the distinctive ways turn exchange may be managed across cultures. Even less well understood, is how human-machine turn exchange is accomplished in interactional contexts where multiple potential communicative modalities are at play. Much more attention has been paid to how systems use sound (Brewster 1998; Rinott 2008), recognize and produce speech that invokes human emotion (Busso et al. 2004; Cahn 1990; Mor 2014; Oakley et al. 2000), and more recently, operate in a multimodal capacity (see Dumas et al. 2009; or Wechsung 2014, for a review).

This piece seeks to address this gap through an analysis of trouble in turn exchange in human-computer multimodal interaction in an in-car infotainment system. One central question we examine is how, in these interactions, where participant expectations of their interlocutor's competence as a model interactant may not hold, does repair get done, and what does the trouble and its repair tell us about the culturally distinctive ways to do it "right"?

In what follows, we review the concepts and theoretical framework employed in the analysis and research design that produced the data set we analyze here, though the theory and methodology as adapted for the study of in-car communication is more fully detailed elsewhere (Carbaugh et al. 2012).

Next, we analyze a number of instances to determine the source of trouble experienced by many participants in interaction with the system, and the variety of methods users employed for accomplishing their goals despite this trouble. Based on this analysis, we highlight cultural norms and premises governing interaction in this communication situation (Hymes 1972). We conclude with a discussion of the implications for multimodal system design.

7.2 Theoretical framework and related literature

In the tradition of Conversation Analysis (see Heritage (2010), for a summary of principles), it has long been accepted that conversation occurs in a sequential fashion, organized through the managed exchange of turns at talk, though debate exists over seemingly contradictory cases (Reisman 1974; Sidnell 2001). Generally, this organization is taken to be fundamental to meaning-making in interaction. Schegloff (2000) captures this stance in the following:

> The orderly distribution of opportunities to participate in social interaction is one of the most fundamental preconditions for viable social organization. ... One feature that underlies the orderly distribution of opportunities to participate in conversation, and of virtually all forms of talk-in-interaction that have been subjected to disciplined empirical investigation, is a turn-taking organization. The absence of such an organization would subvert the possibility of stable trajectories of action and responsive action through which goal-oriented projects can be launched and pursued through talk in interaction ... (p. 1)

The preponderance of articles on the topic of turn-based organization preclude a thorough review here, but for the seminal work of Sacks, Schegloff and Jefferson (1974). Therein, the authors propose a Turn Constructional Unit (TCU) and a Turn Allocation Component. The idea of the TCU suggests that interactants' turns are constructed in such a way as to make the kind of turn it is, the action the turn seeks to accomplish, available to fellow interlocutors such that the projection of the coming end of the turn can be anticipated. That such a function exists in conversation is evidenced by the ability of interlocutors to cut-in before a turn is fully completed, having projected what the completed utterance may likely have contained. The Turn Allocation Component suggests that interlocutors actively manage the exchange and allocation of turns at talk through a variety of practices that are designed to select a next speaker, or self-select as next speaker, and signal when a speaker's turn is completed, or about to be, through a transition-relevance place (TRP). Schegloff (1992) conceptualizes TRPs as "discrete places in the developing course of a speaker's talk (...) at which ending the turn or continuing it, transfer of the turn or its retention become relevant" (p. 116).

The sequential organization of speaking turns also provides the foundation for the interpretation of the meaning of talk in interaction. For instance, returning to our hypothetical conversation between friends in the introduction, if "what concert do you want to see?" is the first turn in a conversation, its meaning may be heard as an invitation to a concert. However, following a prior turn where a friend asks "do you want to see a movie?" that same speech may now be interpreted as indication of a mishearing of the prior turn, or a rejection of the invitation to a movie and a counter-invitation to a concert. Since meaning is reliant on the position of an utterance in relation to surrounding utterances, the timing of conversational turns becomes significant in managing the mutual intelligibility of the interaction, without which the interaction cannot continue without repair. How humans and machines in interaction manage this exchange of turns, and the careful timing required to do so in order to assure mutual intelligibility is of primary concern to this work.

One way interactants can signal that a turn is complete is the use of pause (Maynard 1989). An interlocutor may use a pause at the end of an utterance to invite another speaker's participation, but they may also simply be pausing for breath, to develop their next utterance, because they were distracted, etc. The trouble, then, is determining whether a gap in speech is an invitation to exchange speaking turns, or merely a period of silence where the speaker intends to maintain the floor. Among other strategies, like audible in-breaths, or other disfluencies (Corley & Stewart 2008) such as "umm" or "annnd," culturally competent interlocutors come to know the length of time an interactant might pause to signal a TRP and actively monitor for these in conversation, though syntax (Sacks 1974), prosody (Couper-Kuhlen & Selting 1996), pointing (Mondada 2007) and other pragmatic information (Ford & Thompson 1996) are also potential cues of the coming completion of a turn.

Indeed, in instances of intercultural interaction differences in the use of silence can often create trouble as speakers evaluate the meaning of silence from a cultural vantage. Carbaugh (2005), documents a moment of such trouble in an introductory meeting with a future colleague in Finland, Jussi Virtanen. To Carbaugh's surprise, Virtanen would respond to each of Carbaugh's turns at talk with a 10–20 second pause. From the vantage of an American English speaker from the Northeast, these pauses were exceptionally long and, for Carbaugh, signaled something untoward in the interaction. As he later discovered, the pauses were the result of a confluence of factors including a Finnish customary practice of long (from the American view) pauses after sentences, Virtanen's personal use of longer pauses (from the Finnish view), Virtanen's careful use of English as a second language, and finally the use of long pauses as a means of signifying the respect one has for the occasion and its significance. Here, then, inter-turn pause is both the result of situational factors, but also a motivated use of cultural means for communicating respect and appreciation.

Scollon and Scollon (1981) note trouble with culturally distinctive pause lengths in conversation in their study of interaction between Athabaskan-English speakers and US American-English speakers. The authors find that in conversation, Athabaskans are often overrun by English speakers because of a preference for pausing between the exchange of speaking turns for about a half second longer than typical for US English speakers. This means that English speakers often set the topic of conversation, and then proceed to dominate (from the Athabaskan view) the remainder of the conversation as Athabaskans monitor for TRP's at the end of the English speaker's turn, only to find that the English speaker has started speaking again before they had a chance to take their turn. This leads to negative evaluations of the conversation from both the US English and Athabaskan-English speaker's view, based primarily on cultural variation in inter-turn pause length and the meaning of pauses that last relatively longer or shorter.

Because turn-taking practices serve as the foundation for mutual intelligibility in conversation, and these practices are subject to cultural variation (Tannen 2012), understanding the cultural norms and premises governing the exchange of turns at talk is essential. Attending to issues of variation in norms for turn taking can help illuminate trouble in human-computer interaction in the same way such an analysis can illuminate trouble between members of different cultures. As a result, in the analysis that follows, we make use of concepts from Conversation Analysis, reviewed above, as well as a framework for analyzing the culturally distinctive ways talk is patterned in interaction – Cultural Discourse Analysis (Berry 2009; Carbaugh 1988, 2007, 2012; Scollo 2011).

Cultural Discourse Analysis (CuDA) is a development of the Ethnography of Communication (Hymes 1962, 1972, 1974), that seeks to describe, interpret, compare and critique culturally patterned communication practices. It does so first through the conceptualization of communicative phenomena as a communication "act", "event" or "situation". Communication "acts" are the smallest unit of analysis and may include a single utterance or turn at talk. Communication "events" are bound by a clear initial act and closing act, and contain a structured sequence of acts that are "directly governed by rules or norms for the use of speech" (Hymes 1974, p. 52). The communication "situation" is not bound by particular communication acts as in the communication event, but rather by other boundaries including spatial, such that we might conceive of communication "on the front porch," or "in the corner bar" (Philipsen 1992) as bounded situations that inform the kind of communication in that place. Thus, here, we conceptualize communication in the car as a communication situation, containing a number of communication events, and made up of a number of communication acts.

So conceived, CuDA invites us to investigate these communication acts, events, and situations for "radiants of meaning" found in messages about personhood, social relations, emotion, place, and communication itself. Presumed and enacted in these messages are "cultural premises" which serve as a resource for the interpretation and production of meaning in interaction. Cultural norms may also be identified, which are implicit or explicit rules that govern the moral domain of social action. In our analysis, we employ these concepts in understanding the distinctive premises and norms that both system designers, and users of systems, may employ as they seek to accomplish their respective goals for the interaction, as well as the ways they may sometimes be misaligned, the consequences of that misalignment, and what users do to get back on track.

7.3 Methodology

As discussed in previous work (Carbaugh et al. 2012, 2013; Molina-Markham et al. 2014, 2015) data for the analysis below were collected from the driving sessions of 26 (14 female, 12 male) participants during the study of an in-car infotainment system conducted in Western Massachusetts. During the driving session, participants would use their own car, which had been outfitted with a prototype infotainment system that they would interact with through a dashboard mounted tablet computer. Participants were asked to drive on mostly rural roads of their choosing for one and a half to two hours on average. During this time participants were invited to make use of a variety of the voice capabilities of the system as they would in the normal operation of their own vehicle were they to have such a system. Touch interaction with the system was permitted for starting or ending interactions through use of the microphone button, which started a voice command event, or the "end" button, which could be used to close an ongoing action.

Two researchers were present in the vehicle at all times, one in the passenger seat who made observations and conducted interviews at predetermined stop intervals, and who was ready to take control of the vehicle in case of an emergency. The other researcher was seated in the back seat of the vehicle serving the role of "Wizard," using a laptop to interact with the system to fulfill user directives. Since some of the voice interaction functionality we wanted to test is not yet in production, the human Wizard served the role of the "brain" of the system, interpreting participant's speech and executing directives they had made to the system. We have found no evidence that participants realized the researcher in the back was operating the system.

At the beginning of the driving session participants were instructed to explore the system's multimodal abilities through both touch and voice interaction. After participants felt comfortable using the system, we proceeded to an off-road test where participants drove around a parking lot and further explored voice interaction with the system. Following this we proceeded to an on-road test where participants were instructed to do their best to ignore our presence and use the system as they normally would, which participants generally seemed able to do. At the midpoint of the drive, we conducted a short interview to hear about their experience of the system, answer questions, and suggest functionality they may not have explored or been aware was possible. A similar interview was conducted after the driving session had completed.

Being interested in the sequential organization of multimodal interaction in this context, the way users managed turn exchange with the system, and any potential misalignment between the system's behavior and users' cultural norms and premises operating in this communication situation, we identified interactional sequences where the system's talk overlapped with the user as a sign of potential turn exchange trouble. We found that when users engage in a task-switching event (asking the system to do a task that is not part of the current task the system is performing) a disproportionate amount of overlapping talk occurred relative to other interactional sequences, like directing the system to perform a new task. Thus, the data for this analysis are taken from overlapping talk that occurred during user attempts to switch tasks. Not all users made use of the task switching functionality of the system, which produced a corpus of 7 participants that did, each of whom experienced some degree of overlapping talk with the system on their first use of the task-switch event, making this a regular and robust phenomena for investigation.

7.4 Prompt timing and misalignment – A formula for interruptions

When the system was engaged in the ongoing performance of a given task (playing the radio, making a phone call) and the user pressed the microphone button to initiate a voice command, the system was programmed to respond by providing an audible "ding" sound to confirm that the system received the users request to initiate a voice command and was now in an "on" state. The system would then play a task-relevant prompt. We can already see at play a number of interactional modalities users are negotiating in the opening of this interaction, including touch, audible non-speech in the form of the "ding", and speech from both the system and user, making this a rich and complex interactional context.

For instance, while in the radio task, if the microphone button is pressed, the system would ding and then ask "which station or channel do you want to hear?" It was then the user's turn to talk. However, the sequence never went according to design in the first instance. Below is an example of a typical way this interaction occurred.

Instance 1: Context FM Radio (Participant 11–51:44 minutes into the session).

```
1   P:   (participant touches microphone button)
2   S:   (audible ding)
3        (0.6)
4   P:   phone ca[ll
5   S:          [which station or channel do you want to hear?
6   P:   phone call
```

In this instance, the participant presses the microphone button while listening to an FM radio station. The system responds to the participant's touch with an audible ding. There is then 0.6 seconds of silence before the participant begins her directive to the system – "phone call." During this directive the system overlaps her talk with its own prompt "which station or channel do you want to hear?" The participant responds by restating her directive from the prior turn, "phone call," on line 6, which can be understood as a corrective action since the system's turn was not responsive to her command "phone call". In this way, the system violates Grice's maxim of relevance, issuing an utterance that is not sequentially relevant to the participant's prior turn. In this instance, the user opts to treat the system's violation as the result of a mishearing, and reissues her command.

Why would the system respond in a way that is so badly unresponsive to the participant's directive? At least part of this misalignment, we suggest, is the result of differing meanings for the audible ding the system plays on line 2, and implicates the difficulty of designing multimodal interfaces wherein the meaning or function of certain modes (an audible ding) is not well established. The system is designed to play a ding, followed by a task-relevant prompt, "which station or channel do you want to hear?" The ding, then, from the system's design functions as an acknowledgement of the participant's request to initiate a voice command, a "wake up" chime. The system, however, is designed to take a turn following this ding, verbally prompting the user to provide information relevant to the ongoing task at hand. It plays this verbal prompt, in this instance, 1.5 seconds after the audible ding. The system never manages to play the verbal prompt faster than 1.3 seconds after the audible ding, leaving a sizeable pause between ding and prompt.

Because the system is designed to take the first turn, it is not listening for the participant's voice between the audible ding and its first turn, the verbal prompt. Therefore it cannot move to cut-off its turn in recognition of the participant-issued directive as a human interlocutor might (Scheglof 2000). Because the participant waits only 0.6 seconds before beginning their turn, they issue a command that is not heard by the system, and which causes the system's turn (its first turn from the system's view) to be badly non-responsive to the sequential position of the interaction at that juncture.

Now that we understand what the system thinks is happening we might ask, why does the participant take their turn at 0.6 seconds? One possible explanation is misalignment between the system and participant on the meaning of the audible ding on line 2. The participant may understand this ding in a number of ways. We first suggest that the participant might understand the ding as a summons response, borrowing from the interactional form of the telephone conversation.

In 1964, Sacks (Jefferson & Scheglof 1995) pioneered studies of telephone call interactions and concluded that the ringing telephone functions as a summons and the answering of the phone with "hello" functions as a response to the summons (Sacks 1974), ostensibly two turns at talk have been exchanged. The next turn, then, wherein the topic of the conversation is set, belongs to the actor who did the summoning, the caller. In this case, the participant issues the summons through touching the microphone button, and the system responds with an audible ding. If this interaction were following the routine form of the telephone call then the participant would take the next turn and set the topic. Instead, here, the system attempts to set the topic by asking what station the participant wants to hear. It is possible, then, that participants are modeling interaction with the system after the routine form of the telephone call, wherein the participant takes the first turn at talk, and that this understanding informs their move to initiate their directive before the system's verbal prompt, since they do not expect the system to be taking a turn in this position.

Also at work are cultural norms for the amount of time that passes in a gap between turns before that gap signifies a Transition Relevance Place (TRP) where a conversational turn may be understood to be over or relinquished. If a cultural norm exists for the participant that turns are exchanged after roughly 0.6 seconds of silence, then the system will routinely take too long to take its first turn as users proceed to interpret the system's silence as yielding the speaking floor. As we see in the following instances this appears to be the case.

Instance 2: Context My Music (P8 – 28:06).

```
1   P:   (Participant touches microphone button)
2   S:   (audible ding)
3        (1.0)
4   P:   next
5   S:   what artist would you like?
6   P:   next
```

In this instance, like the last, the user is in a task, in this case listening to their downloaded music library, when she decides to touch the microphone button. After doing so the system dings, and after a 1 second gap she issues her directive "next." The system's next turn, which sequentially would be heard as a reply to the participant's directive, asks the participant what artist she would like to hear. This verbal prompt is of course not responsive to the participant's directive, and so the participant restates it on line 6, again treating the system's turn as the result of a mishearing in need of correction. The amount of time it takes the system to respond to the microphone button press was measured at 2 seconds, this is the longest the system takes to issue a prompt. This allows the participant to wait 1 second and then issue her directive in the remaining 1 second before the system plays its prompt. Because of this the user does not experience overlapping talk with the system, but is perhaps presented with an even more confusing response, since the overlap itself can function to let the participant know that something is wrong. Without the benefit of the overlap, the user is left to wonder what the system's turn means and what should be done next?

This 1 second pause, like the 0.6 second pause in the prior instance, is long enough to indicate to the participant that the system has yielded the floor, and it is now her turn. This is not the case however, from the system's design, and the system proceeds to issue what it takes to be its first turn, leaving the participant to conclude that the system has either not heard, or misheard her command. This may negatively impact participant perception of the competence of the system as a voice interaction partner. We can see the persistence of this pattern in Instance 3 below.

Instance 3: Context XM Radio (P12 – 23:13).

```
1   P:   (Participant touches microphone button)
2   S:   (audible ding)
3        (1.5)
4   P:   ca[ll
5   S:     [what xm channel?
6        (1.7)
7   P:   call tom
```

In this instance, the participant is listening to an XM radio station when she presses the microphone button. The system responds to the press with an audible ding, at which point the participant waits 1.5 seconds before beginning to issue her directive "call." However, as she begins to issue this directive, the system overlaps her speech with its own question, "what xm channel?" After this, the participant waits 1.7 seconds and then restates the directive she appeared to be beginning on line 4.

The participant in this instance treats 1.5 seconds as sufficient time to signify a TRP, inviting her to take her turn. However, here, the participant has waited long enough that the system begins its prompt almost simultaneously with the start of her directive. Unlike the prior instances the participant here abandons her turn, surrendering the floor to the system in an overlap resolved after one beat (Scheglof 2010). Overlaps which do not end after one beat may be the beginning of an indication of "competitive production" wherein two or more interlocutors vie for the floor in a competitive move. One can imagine that the ideal design of the system would not be such as to enter into competition with its users for turns at talk, being more preferably oriented to user-satisfaction and compliance. However, here, the system produces a turn wherein the user is forced to either competitively co-produce talk until the system completes its turn, or abandon their turn thereby deferring to the system's "right to speak". This is open to being heard by participants as a dominant interactional move, which is likely not a desirable position for the system and user.

Across these last three instances, participants encountered the same interactional trouble, attempting to initiate a directive to the system that the system responds to with a sequentially non-responsive verbal prompt, and/or in most cases, the participant's talk is overlapped by the system's forcing the participant to compete for the floor or abandon the turn.

Participants in the larger corpus from which these instances are taken varied in the amount of time they waited before speaking after the audible ding from the system from 0.6 seconds to 1.5 seconds. Since the average time the system takes to generate a prompt following the audible ding is 1.7 seconds, this means those participants who wait around 1.5 seconds to give the system a directive will almost certainly be interrupted by the system, while those who begin a directive immediately following the ding may be able to complete their directive utterance, only to be met with a question that seems irrelevant to the directive they have issued. A seemingly simple fix for this trouble is an anti-overlap feature to assure that if the system hears the user talking, it hold its turn until the system can decide what next action to take that would be relevant to the user's speech. However, listening for user speech all the time when it has no reasonable expectation that the user is about to speak, like after a TRP, means lots of mistaken "hearings" on the system's part that could lead to even more trouble.

The patterning of the interaction above is suggestive of a norm for the management of turn exchange in conversational positions where turn allocation is ambiguous. This norm treats pauses of longer than 0.6 seconds, and no longer than 1.5 seconds, to be indicative of the passing of a turn. The system's routine pause length of 1.3–2 seconds, then creates a misalignment in the turn-taking management of interaction between the participant and system. As a result, participants are forced to abandon their turn, or competitively produce a turn in overlap with the system. Participants must further make sense of the system's turn, which given its late positioning in the interaction relative to the position it was designed to inhabit, appears non-responsive to the participant's directive. In the instances above, participants treated the system's turn as a mishearing in need of correction through repetition of the initial directive, though this was not the only way participants managed to negotiate the difficulty of this misalignment. After having encountered this misalignment some number of times participants would generally adjust their interaction with the system in one of four ways, which we review in the following section.

7.5 Interactional adaptation

After a participant experienced the system overlapping their directive, and/or responding to their directive in non-responsive ways, they appeared to learn, at different rates, that the system will be taking a turn after the audible ding in task-switch events, and that this turn will take place after some notable pause. Given their apparent noticing that this is the case, users proceeded in future interactions with the system in one of four ways.

(1) One way participants adapted to the system was to sustain a "competitive production" (Schegloff 2000). In the instance below this is accomplished through the extra-ordinary elongation of the vowel sound in "too," sustained until after the system's turn had completed.

Instance 4: Context FM Radio (P10 – 1:24:33).

```
1   P:   (Participant touches microphone button)
2   S:   (audible ding)
3        (0.8)
4   P:   change st[ation too::::::::::::::::::::::::::::::::::::owuh (1.2)
5   S:            [what radio station do you want to hear?
6   P:   ninety seven point three
7        (4.5)
8   S:   could you repeat that please
```

In this case, the participant, unlike the participant in instance 3, refuses to abandon their turn to the system and makes a bid to hold the floor through the elongation of the vowel sound in "too" on line 4. A fellow human interactant would then be forced to choose to continue their own turn in a sustained overlap, or yield the floor to the other speaker. Because the system is not listening when it's playing its own prompt, it is incapable of knowing that the participant is speaking and therefore incapable of deciding to abandon its turn. This means the system, in instances of competitive production, will always sustain overlap until its turn is complete. Somewhat ironically, however, the system can never "win" since human interactants engaged in competitive production can project the incipient end of a turn shape and adjust their strategy for elongating their turn to assure it lasts longer than the system's. This is the case in the instance above.

Despite the participant "winning" the competitive production, the directive to tune to 97.3 cannot be understood by the system because it was not listening to the participant's utterance during the overlap. Even if the system had been listening it would not be able to understand a directive including the extraordinary stretched vowel seen here. A bitter-sweet victory. The implications for the outcome of this competition among human interactants would likely include messages about the status of the relationship between interactants. As Tannen (1993) points out, however, the meaning of this overlap to participants is not set *a priori* as Schegloff's (2000) use of the term "competitive production" might suggest. Overlap may also be understood by interactants as a move to solidarity, though it does appear in this instance that the participant intends to outlast the system. Regardless, the possible interpretations of the meaning of the overlap to participants, one thing is certain, the system will take no implications about social relations from the interaction, though the participant may.

This is one way participants have borrowed interactional strategies (competitive production) from human interaction for use in dealing with an invasive conversational partner (the system) but where the social effects of which may not carry over. This seems to not discourage their use here however.

(2) The second way users adapted to the system's overlap was to wait out the long pause for the prompt and then speak. In this strategy, participants allow for a longer pause than they generally had in the past, giving the system the opportunity to play its prompt. The participant would then give their directive, which was necessarily shaped as a corrective, since quite often the directive they gave the system was not related to the task-specific prompt the system played.

Instance 5: Context My Music (P8 – 31:08).

```
1   P:   (Participant touches microphone button)
2   S:   (audible ding)
3        (1.3)
4   S:   What artist would you like?
5   P:   FM Radio
6   S:   Just a second
```

Here the user touches the microphone button and the system dings in reply, 1.3 seconds pass, and the system asks the participant what artist she would like to hear. The participant, apparently not wanting to hear an artist replies "FM

Radio". Structurally, the exchange of turns has gone smoothly here (no overlap) though there are two things to note. First, this participant experienced trouble with a turn exchange performing a task switch 4 minutes prior (Instance 2), experiencing the overlap phenomenon common among all users. As a result, we suggest that her decision to not take a turn during the 1.3 seconds gap after the system's "ding" is an adaptation to the overlap trouble from her prior task-switch experience. This, then, is another way users have developed to deal with turn-exchange difficulty, as ultimately the participant has goals for the interaction she would like accomplished and needs to find a way to get back on track in order to do so. In this case, this is accomplished through the participant's adaptation to the system's norm for a 1.3–2 second pause before its first turn. This participant has then moved through a process to identity the trouble (the system intends to take a turn at talk after the ding, and has a relatively long pause before it does so), develop a possible solution (wait until the system speaks) and implement that solution, though it contradicts her and other participants' routine norm for managing turn exchange (a 0.6–1.5 second pause).

Despite having adjusted the timing of her initial turn to accommodate the system, she is still placed in the position of having to respond to the system's prompt with a move to reject the system's offer. Since the system opts to take a guess at what the participant might want, asking "what artist would you like?" the system constructs its turn as to prefer a response that chooses an artist. Any response from the participant that is not the name of an artist is thereby shaped as a dis-preferred response (Heritage 1983; Levinson 1983; Pomerantz 1984). This restricts the available next actions for the participant to either a response to the question that selects an artist, or an outright rejection of the system's offer to play an artist, which people would generally rather not have to do.

Another participant adopted this same strategy, also 5 minutes after experiencing an overlap with the system during a task-switch event.

Instance 6: Context XM Radio (P12 – 28:45).

```
1   P:   (Participant touches microphone button)
2   S:   (audible ding)
3        (1.3)
4   S:   what XM channel?
5   P:   bridge
6        (7.5)
7   S:   hold on
```

Given the proximity of this instance to the participant's prior system overlap, it is likely here that the user has adjusted her turn timing to accommodate what she now knows to be the system's long pause following the audible ding. As it happens in this sequence, allowing the system to take its turn after the extended pause provides the system a chance to guess that the user might want another XM channel. Often this guess is wrong, as in the last instance, which leads to the necessity of a participant's rejection of the system's guess, but in this case the guess is right as the user does not wish to switch tasks, only channels. This is, then, the best case scenario, though it requires a deviation from the user's established interactional norms for pause that indicates a TRP in order to accomplish.

(3) Another option participants developed for dealing with the overlap involved an abandonment of the use of the task switch capacity. In this sequence, participants opted to switch tasks by first touching the "End" button to stop the current task (playing the radio), and then pressed the microphone button to initiate a new voice command. When the microphone button is pressed outside of an ongoing task, like when the user is on the system's home screen, the system plays an audible ding and then waits for the participant's command. Possible overlap is then avoided by selecting an interactional path that does not include the system taking a spoken turn. One benefit, then, of multimodal systems is the ability of the user to adapt to verbal interactional trouble by employing alternate modes that avoid the trouble.

Instance 7: Context XM Radio (P9 – 53:05).

```
1   P:   (Participant touches End Radio button)
2        (2.0)
3   S:   (Radio stops playing, screen shifts to home)
4        (.5)
5   P:   (hand begins move toward radio)
6   P:   (.7)
7   P:   (finger touches mic button)
8   P:   gimme dubbelyu=efem (.) ehhhn give me doubelyu::::::
             whatsitcalled effseear
```

Here the user begins by ending an ongoing task, the playing of the radio by touching the End Radio button. It takes the system 2 seconds to comply with the user's directive to stop playing the radio and return to the home screen on the display. Within 0.5 seconds the user begins to move his hand back toward the radio and makes contact with the microphone button 0.7 seconds later. He then gives the radio a directive to play WFCR, all in less time than it took the system to comply with his initial directive to end the radio.

It is likely, then, that the user intended to change radio stations when he hit the end radio button, but why not just press the microphone button and tell the system to change stations, making use of the system's task-switch function? The answer we propose that best accounts for the participant's actions here is that in prior interactions the user had difficulty with the turn exchange, particularly, during a task switch event 40 minutes prior, whereafter he ceased to use the task-switch functionality, opting instead to explicitly end all ongoing tasks through touch before initiating a voice interaction to issue a new command. Abandoning the line of interaction that produced the turn-exchange difficulty is then one method a user developed for accomplishing the task sought, despite trouble with the timing of turn exchanges in the task-switch event. Doing so is likely not the ideal case however, as the task switch function allows users to achieve their goal in as little as one button press and one voice command, while the strategy adopted by this participant will require a minimum of two button presses and a voice command. This is not ideal from the perspective of system designers either, since the minimization of the use of touch while driving is preferred for safety reasons.

(4) Not all users did develop new methods for dealing with the overlap trouble. One participant continued to repeat the pattern observed in instances 1–3 (issue command, system overlaps, reissue command) 6 times repeatedly, one after the other, over the course of her drive with the first occurrence at minute 20 and the last 42 minutes later. During this time the participant never adjusted the pattern of their interaction, continually experiencing overlap with the system each time she performed a task switch. We have included one of these instances here for illustration, though the patterning is identical to instances 1–3 reviewed above. The following is taken from the fourth recurrence of this pattern with this participant.

Instance 8: Context FM Radio (P11 – 54:19).

```
1   P:   (Participant touches microphone button)
2   S:   (audible ding)
3   P:   phone call
4   S:   which station or channel do you want to hear
5   P:   phone call↑
6   S:   okay (1) who would you like to call
```

That this participant persists across a number of instances to issue a directive prior to the system's turn is likely the result of the participant never identifying that the system intends to take the first turn at verbal interaction and as a result is not listening as it prepares its turn. Instead, the participant through the repetition of her initial directive, treats the system's prompt as a mishearing of her initial directive in need of repetition. If the participant had identified that the system was not listening, persisting with issuing the command before the system's prompt would serve no purpose and would likely have discontinued. This is suggestive that not all participants have equal access to the resources required to interpret the source of trouble with the system's behavior. This instance further highlights the trouble of poorly crystalized norms surrounding the meaning and turn status of the audible "ding" in multimodal interaction.

It is also possible that if the participant was not able to ultimately accomplish the task she sought as a result of this trouble then she may have proceeded to explore other strategies. However, in this instance, the participant is ultimately able to get the system to follow her directive on line 6 when the system acknowledges her directive to make a phone call asking "who would you like to call." This participant then appears willing to accept some trouble so long as the task is accomplished in the end.

7.6 Norms and premises

The analysis above suggests some normative ways that participants approach interacting with the system, as well as certain premises that inform this use. In the instances presented above, each participant experienced either an overlap of talk with the system and/or a seemingly non-responsive reply to their directive. The regularity with which this phenomenon occurred throughout the corpus suggests that the amount of time these participants understand as evidence of, or opportunity for, a turn exchange, in contexts where next speaker is ambiguous, is less than the 1.7 second average time the system takes to produce its verbal prompt. We believe this amount of time to be a cultural norm for managing the interactional exchange of conversational turns when the next speaker is ambiguous. The system, then, is engaged in a kind of norm violation when it produces its overlapping prompt that carries the interactional force of an interruption, violating the moral order of turn-taking and politeness that is generally expected between human interactants in social interaction. This norm can be more explicitly formulated as: *In contexts where next speaker is ambiguous, if an interactant wishes to take a turn, they should do so between 0.6 and 1.3 seconds after the prior action, in order to be a proper interactant.*

Misalignment between the system and participants was not restricted solely to the normative timing of the exchange of speaking turns, but also in the meaning of particular actions within a communication event, as in the case of the audible ding. Whereas the system was designed with the audible ding's intended meaning being an alert of the system's status, akin to announcing the system is on, participants treated the audible ding as a summons response akin to the organization of telephone calls. Depending on which meaning of the audible ding one employed, a different next speaker would be appropriate. From the participants' vantage the audible ding occupied the space of an interactional turn, and therefore the system was understood to have passed the turn back to the participant for their first spoken turn.

In all but one case, participants chose to adjust their interactional strategies for accomplishing the task they sought, with one user persisting in the original pattern of overlap, likely doing so as the result of failing to identify that the system was not listening in the gap after the audible ding. This means that all participants who became aware of the source of the trouble opted to make adjustments in order to accomplish the task.

It is not automatically the case that this should be so. In human interaction, an interlocutor behaving in the way the system does would likely be called to account for both their repeated interruptions, but also for violations of the maxim of relevance for no apparent conversational purpose. However, the participants in the instances collected above never call the system to account for its behavior, nor exhibit any animus toward the system for what might be cause for an argument with a human interactant. The system, in effect, gets a pass. This is not to say participants will, or do, find interacting with a system under these conditions pleasing, only that the system itself appears not to be held responsible for its behavior in this context. This is likely because participants understand that the system lacks the fully fledged capabilities of a culturally competent human interactant, and therefore cannot be held responsible for these sorts of issues, but likely not trusted either.

A premise of and for communication can then be identified in the participants' interaction that *those who are not fully competent interactional partners cannot be held responsible for certain interactional blunders.* An accompanying premise of personhood can then also be formulated as *voice interactive machines are not fully competent interactional partners.* And finally, an additional norm can be identified for proper behavior given the above premises, *since machines are not fully competent interactional partners, human users ought to adjust to the system in order to accomplish their goals.* These premises likely inform the level of tolerance users have for interacting with systems that routinely violate human interactional norms, without which interaction with systems at this level of capability would not be possible. This does not mean that the above premises are universal or automatic, as one can imagine alternate premises that systems such as these can be held responsible for interactional blunders, such that repeated violations of the interactional order result in discontinued use of the system. This did not, however, appear to be the case in participant interaction with the system during task-switch events in this corpus. It is unclear whether the research context in which these data were collected resulted in more persistent attempts to continue using the system than might have occurred had the user been alone in their own vehicle.

7.7 Implications for design

The analysis above can be used to make particular recommendations for the improvement of multimodal interactive systems in the future. First, the task-relevant prompt is problematic as many participants noted in interviews that it was unnecessary, inappropriate, or too long. Some participants suggested that no prompt was needed at this stage of the interaction at all, citing that when someone presses the microphone button while in a task they likely have something they would like to do in mind, and have pressed the microphone button in order to give the system that command. As a result, the system need not offer any prompt, but rather just listen for the participant's command.

Second, system designers need better understand the role of non-speech sounds in multimodal interaction and their turn taking relation to other modalities such as touch and speech. In the design of this system, the audible ding is treated as if it occupies no conversational position – it takes no turn. This is clearly not how participants in the above interactions understand the ding. A turn-based analysis of the interaction suggests that the ding does function as a turn-at-talk, with the first turn being the participant's touch of the microphone button, the second turn being the system's reply to the touch through audible ding, and the third turn then passing back to the user for first topic. However, because the system design does not account for the audible ding as an interactional turn, it presumes users will wait for the system to respond to the microphone press with a verbal prompt. This appears to not be the case as users hear the audible ding as the response to the microphone press and proceed to take their turn. Some research on the role of non-speech sounds in human-computer interaction is already underway (Brewster 1997; Brewster 2002; Hereford & Winn 1994), but does not incorporate an analysis of the sequential position of this mode in the organization of interaction.

Some participants did report a desire to have a system prompt after the microphone button was pressed as a sign that the system is "listening," but thought the prompt that was offered was simply too long for regular use. Participants suggested alternate prompts including "yes?" or "what would you like?" which are likely better alternatives as they are task independent and do not require users to reject the system's wrong guess, which as indicated above is a dispreferred action in conversation.

Ultimately, however, what holds multimodal interactive systems back the most in the instances analyzed above is the system's inability to listen for user speech and act accordingly. Cases of overlap in human interaction are resolved in a variety of ways (Schegloff 2000) but all require monitoring of the ongoing turn by both interactants. In order to properly model human interaction, the system must be able to listen to users' ongoing turns and adapt, as we do with them. Research on the broader phenomena of overlapping speech in HCI, sometimes referred to as "barge-in," is also underway, examining the frequency and context of "barge-in" cross-culturally (Wang, Winter & Grost 2015)

We further advocate attention be paid to the cultural nature of the management of turn-exchange both in the amount of time interlocutors normatively wait as indication of a TRP, but also in the practices employed managing turn exchange, and the strategies adopted to accomplish interactants' goals. The analysis above suggests two cultural norms and two premises of and for communication and personhood that may vary culturally and influence the way users interact with these kinds of systems, particularly surrounding the resolution of trouble and the meaning of that trouble.

Abbreviations

FM   (radio)
MP   Modal Person
TCU   Turn Constructional Unit
TRP   Transition-relevance Place
CuDA   Cultural Discourse Analysis
XM   (radio)
WFCR   (radio station)
HCI   Human Computer Interaction

References

Berry, M 2009, 'The social and cultural realization of diversity: An interview with Donal Carbaugh', *Language and Intercultural Communication*, vol. 9, pp. 230–241.
Bossemeyer, RW & Schwab, EC 1990, 'Automated alternate billing services at Ameritech', *Journal of the American Voice I/O Society*, vol. 7, pp. 47–53.
Brewster, SA 1997, 'Using non-speech sound to overcome information overload', *Displays*, vol. 12, no. 3–4, pp. 179–189.
Brewster, SA 1998, 'Using non-speech sounds to provide navigation cues', *ACM Transactions on Computer-Human Interaction*, vol. 5, no. 2, pp 224–259.

Brewster, SA 2002, 'Nonspeech auditory output', in *The human-computer interaction hand-book: Fundamentals, evolving technologies*, eds A Sears & J Jacko, CRC Press, pp. 221–237.

Brown, P & Levinson, SC 1987, *Politeness: Some universals in language usage*. Cambridge University Press, Cambridge.

Busso, C, Deng, Z, Yildirim, S, Bulut, M, Lee, CM, Kazemzadeh, A, Lee, S, Neumann, U, & Narayanan, S 2004, 'Analysis of emotion recognition using facial expressions, speech and multimodal information', *Proceedings of the 6th International Conference on Multimodal Interfaces*, State College, PA, USA, pp. 205–211.

Cahn, JE 1990, 'The generation of affect in synthesized speech', *Journal of the American Voice I/O Society*, vol. 8, July, pp. 1–19.

Carbaugh, D 1988, *Talking American: Cultural discourses on DONAHUE*, Ablex, Norwood, NJ.

Carbaugh, D 2007, 'Cultural Discourse Analysis: Communication practices and intercultural encounters', *Journal of Intercultural Communication Research*, vol. 36, no. 3, pp. 167–182.

Carbaugh, D 2012, 'A communication theory of culture'. *Inter/Cultural Communication: Representation and Construction of Culture*, ed A Kurylo, Sage, Thousand Oaks, pp. 69–87.

Carbaugh, D, Molina-Markham, E, van Over, B & Winter, U 2012, 'Using communication research for cultural variability in HMI design', in *Advances in human aspects of road and rail transportation*, ed N Stanton, CRC Press, Boca Raton, FL, pp. 176–185.

Carbaugh, D & Poutiainen, S 2005, 'Silence, and third-party introductions: An American and Finnish dialogue', in *Cultures in conversation*, Lawrence Erlbaum Associates, Mahwah, NJ, pp. 27–38.

Carbaugh, D. Winter, U, van Over, B, Molina-Markham, E & Lie, S 2013, 'Cultural analyses of in-car communication', *Journal of Applied Communication Research*, vol. 41, no. 2, pp. 195–201.

Corley, M, Stewart, OW 2008, 'Hesitation disfluencies in spontaneous speech: The meaning of um', *Language and Linguistics Compass*, vol. 2, no. 4, pp. 589–602.

Couper-Kuhlen, E & Selting, M 1996, 'Towards an interactional perspective on prosody and a prosodic perspective on interaction', in *Prosody in Conversation*, eds Couper-Kuhlen & Selting, Cambridge University Press, Cambridge, pp. 11–56.

Dumas, B, Lalanne, D & Oviatt, S 2009, 'Multimodal interface: A survey of principles, models and frameworks', in *Human Machine Interaction: Lecture Notes in Computer Science*, eds D Lalanne & J Kohlas, pp. 3–26.

Ford, CE & Thompson, SA 1996, 'Interactional units in conversation: Syntactic, intonational, and pragmatic resources for the management of turns', in *Interaction and grammar*, ed EA Schegloff & SA Thompson, Cambridge University Press, CAmbridge, pp. 135–84.

Grice, P 1975, 'Logic and conversation', in *Syntax and semantics: Vol. 3 speech acts*, eds P Cole & JL Moran, Academic Press, New York, pp 41–58.

Hereford, J & Winn, W 1994, 'Non-speech sound in human-computer interaction: A review and design guidelines', *Journal of Education Computing Research*, vol. 11, no. 3, pp. 211–233.

Heritage, J 1983, *Garfinkel and Ethnomethodology*, Polity, Oxford.

Heritage, J 2010, 'Conversation analysis: Practices and methods', in *Qualitative research* (3rd ed), ed D Silverman, Sage, London, pp. 208–230.

Hymes, DH 1972, 'Models of the interaction of language and social life', in *Directions in sociolinguistics: The ethnography of communication*, eds JJ Gumperz & D Hymes, Holt, Rinehart & Winston, New York, pp. 35–71.

Hymes, DH 1974, *Foundations in sociolinguistics: An ethnographic approach*, University of Pennsylvania Press, Philadelphia.

Jefferson, G & Schegloff, E 1995, *Lectures on conversation*, Willey Blackwell.

Levinson, SL 1983, *Conversational structure*, Cambridge University Press, Cambridge.

Maynard, SK 1989, *Japanese conversation: Self contextualization through structure and interactional management*, Ablex, Norwood, NJ.

Molina-Markham, E, van Over, B, Lie, S & Carbaugh, D 2014, '"You can do it baby": Non-task talk with an in-car speech enabled system.' Manuscript submitted for publication.

Molina-Markham, E, van Over, B, Lie, S & Carbaugh, D 2015, '"OK, talk to you later": Practices of ending and switching tasks in interactions with an in-car voice enabled interface', in *Globalizing personas: Employing local strategies research to understand user experience*, ed T Milburn, Lexington Books.

Mondada, L 2007, 'Multimodal resources for turn-taking: pointing and the emergence of possible next speakers', *Discourse Studies*, vol. 9, no. 2, pp. 194–225.

Mor, Y 2014, 'The future of human-machine interaction: It's not what you say, it's how you say it', *Wired*. Retrieved from: http://www.wired.com/2014/02/future-human-machine-interaction-say-say/. [21 February 2014].

Oakley, I, Brewster SA & Gray, PD 2000, 'Communicating with feeling', in *Proceedings of the First Workshop on Haptic Human-Computer Interaction*, pp. 17–21.

Philipsen, G 1992, *Speaking culturally*, State University of New York Press, Albany, New York.

Pomerantz, A 1984, 'Agreeing and disagreeing with assessments: some features of preferred/dispreferred turn shapes', in *Structures of Social Action*, eds JM Atkinson & J Heritage, Cambridge University Press, Cambridge, pp. 57–101.

Reisman, K 1974, 'Contrapuntal conversations in an Antiguan village', in *Explorations in the ethnography of speaking*, eds R Bauman & J Sherzer, Cambridge University Press, Cambridge, pp. 110–124.

Rinott, M 2008, 'The laughing swing: Interacting with non-verbal human voice', *Proceedings of the 14th International Conference on Auditory Display*, Paris, France, June 24–27, 2008.

Sacks, H, Schegloff, EA & Jefferson, G 1974, 'A simplest systematics for the organization of turn-taking for conversation', *Language*, vol. 50, no. 4, pp. 696–735.

Schegloff, EA 1992, 'To Searle on Conversation: A note in return', in *(On) Searle On Conversation*, eds H Parret & J Verschueren, Benjamins, Amsterdam, pp. 113–128.

Schegloff, EA 2000, 'Overlapping talk and the organization of turn-taking for conversation', *Language in Society*, vol. 29, no. 1, pp. 1–63.

Scollo, M 2011, 'Cultural approaches to discourse analysis: A theoretical and methodological conversation with special focus on Donal Carbaugh's Cultural Discourse Theory', *Journal of Multicultural Discourses*, vol. 6, pp. 1–32.

Scollon, R, & Scollon, S, 1981, *Narrative, literacy, and face, in interethnic communication*, Ablex, Norwood, NJ.

Sidnell, J 2001, 'Conversational turn-taking in a Caribbean English creole', *Journal of Pragmatics*, vol. 33, no. 8, pp. 1263–1290.

Stivers, T, Enfield, NJ, Brown, P, Englert, C, Hayashi, M, Heinemann, T, & Levinson, S 2009, 'Universals and cultural variation in turn-taking in conversation', *Proceedings of the National Academy of Sciences*, pp. 106–126.

Stokes, R & Hewitt, J 1976, 'Aligning actions', *American Sociological Review*, no. 41, pp. 46–42.

Tannen, D 1993, 'The relativity of linguistic strategies: Rethinking power and solidarity in gender and dominance', in *Gender & Discourse*, Oxford University Press, New York & Oxford, pp. 19–52.

Tannen, D 2012, 'Turn-taking and intercultural discourse and communication', in *The Handbook of Intercultural Discourse and Communication*, eds CB Paulston, SF Kiesling & ES Rangel, Blackwell Publishing, pp. 135–157.

Wang, P, Winter, U & Grost, T 2015, 'Cross cultural comparison of users' barge-in with the in-vehicle speech system', in *Design, user experience, and usability: Interactive experience design*, Lecture Notes in Computer Science, ed A Marcus, Springer International Publishing, Switzerland, pp. 529-540.

Wechsung, I 2014, *An evaluation framework for multimodal interaction: Determining quality aspects and modality choice*, Springer International Publishing, Switzerland.

Yael Shmueli-Friedland, Ido Zelman, Asaf Degani, and
Ron Asherov

8 Towards objective method in display design

Abstract: With advances in information technologies, the amount and complexity
of information provided to the driver is growing at a rapid pace, making traditional designs of the instrument cluster display insufficient. Concurrent advances
in display technology enable prudent departure from traditional design solutions
and support new forms of information organization in this confined space. This
chapter aims to address the process of information organization in general, with
special emphasis on the visual cluster display. It illustrates a formal approach that
can be used in the analysis (evaluation) as well as the synthesis (creation) processes of a visual display design. The approach relies on the ability to quantify the
relationships between informational elements in the display and identify opportunities for better integration and configuration of these elements using several
analytical steps. It is proposed that this methodology can be generalized to the
problem of sound and haptics as well as multimodal interface design as a whole.

8.1 Introduction

"Instrument cluster" is the technical term for the main display of information located just behind the driver's wheel. It includes the speedometer, tachometer, fuel
information, engine alerts and optionally other information. Over the decades of
motorized transportation, the organization of information in the "cluster" has
to a large extent remained standard, with some minor variations between car
manufacturers. The introduction of driving assistance features such as Adaptive
Cruise Control and Lane Centering, and the general shift from manual driving
to partially-automated driving, require serious reconsideration of information
organization in modern automotive cockpits. The recent advances in display
technology and the introduction of a flat screen display (now replacing the classical wood or plastic dashboard with bored holes for gauges) allow designers to
organize cluster information in any shape, color, and arrangement.

An important factor in the design of modern clusters concerns the availability
of data and information. There is an ever-increasing insertion of computing, processing and analytics power into today's automobiles (Broy et al. 2007; Charette
2009). From a cluster design point of view, this translates into the availability of
large amounts of data and information that can be provided to the driver. Natu-

rally, all these technologies have or will have an immediate and profound impact on the driver's operating environment, as they place increased demands on the driver to understand, monitor and interact with the car systems. This problem becomes acute when we consider the role of information presentation in automated driving situations. The key question, from our perspective here, is how to best manage the wealth of available information and how to best organize it in a correct, efficient, elegant, and pleasing way within a small visual space (Dingus & Hulse 1993).

One approach to the problem of organization of information is to consider it as a hierarchy, starting from a process of abstraction of the data into information (e.g. removal of superfluous and irrelevant data), then integration of separate informational elements into sets (e.g. by focusing on meaningful interrelationships), and finally configuration, or arrangement, of these sets of information into a whole – much like a tile-work is being woven to create a façade (Degani et al, 2009; Shmueli et al. 2013). In this chapter we focus our attention on the problem of integration; namely, how to take advantage of interrelations between individual elements of information, so as to arrange them in a cohesive manner. Analysis of many examples of integration (Alexander 2002a, 2001b; Degani 2014) suggests that finding a "good" solution involves creating a coherent structure of several individual elements. We define integration as the process of molding several informational elements that are practically and/or conceptually related into a coherent structure, in order to reveal and highlight meaningful interrelationships and support quick recognition. In creating coherent structures, each element not only has to stand on its own, but also to support and preferably to enhance other elements (cf. Bennett & Flach 2011, Ch. 11); the stronger the relationships and mutual support, the more potent the design (Alexander 2002a, Ch. 3).

Figure 8.1 (a) shows a generic driver's display, standard in the sense that it includes many of the informational elements that exist in modern cars, as well as a dedicated pictorial region for automation features such as adaptive cruise control (ACC) and lane keeping assist that are beginning to appear in luxury cars. The adaptive cruise control system (Labuhn & Chundrlik 1995), operates like a regular cruise control system (see Fig. 8.1 (b) for a larger view): the driver selects the "set" speed and the car maintains that speed automatically (with no need to press the gas pedal). The adaptive part involves a radar system that detects the speed of a preceding car. When a slower car is detected ahead, the system automatically reduces the *own car speed* so as to maintain the *preceding car's speed* (such that a preset gap distance is preserved). The "set" speed, initially selected by the driver for the cruise control system, is indicated on the speedometer by means of a small green triangle above the speed graduation. When the system detects a slower car ahead, it reduces speed to maintain the gap from the preceding car, and the arc

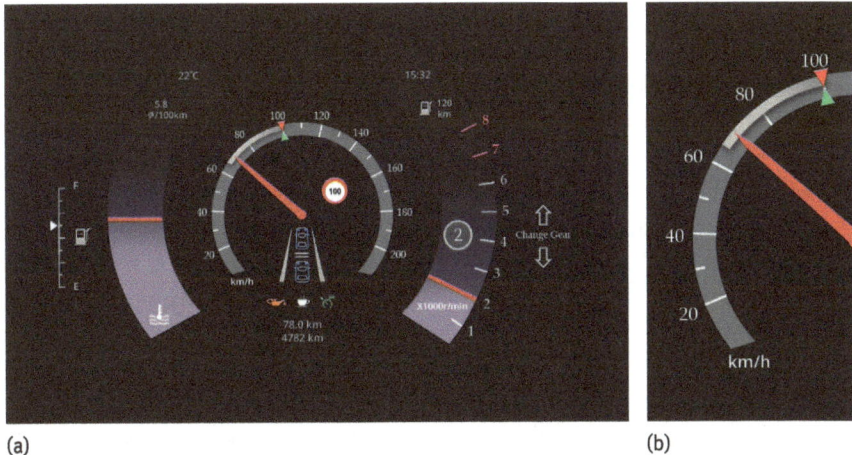

(a)   (b)

Fig. 8.1: (a) A generic instrument cluster. The ACC "set" speed is indicated by a small green triangle. When the system detects a slower car ahead, it reduces speed to maintain the gap from preceding car and the arc between the preceding car's speed and the "set" speed opens up (in light silver). The arc highlights the difference between the preceding car's speed and the set speed. (b) A detailed view of the integrated structure associated with ACC usage.

between the preceding car's speed and the "set" speed appears in silver. This arc highlights the difference between *the preceding car's speed* and *set speed*. These two informational elements are perceptually grouped together to form an integrated unit. A quick glance at this unit can inform the driver about the state and status of the ACC and how much the speed of the preceding car differs from the *set speed*. The speed needle allows the driver to assess the proportional relationship between the *set* and *preceding car's speed* with respect to *own speed*.

It is possible to qualitatively describe the integration of information indicated as follows:

The first element is *ACC set speed* – the speed of the ACC as set by the driver (100 km/h) marked by a small green triangle. The second is the *Preceding car speed* (67 km/h). Additionally, the green triangle and the silver arc indicate to the driver that the ACC system is on. The silver arc alone, indicates that the radar is working, and that there is a car ahead of us. The *own car* is going at a speed of 70 km/h, to accommodate the slower car in front. The emergent feature here, which is a result of the geometric configuration of the integrated unit of information, is the Δ1 speed between *set speed* and *preceding car* speed (100 – 67 = 33 km/h). If we had provided these two pieces of information in an alphanumeric format, the emergent geometrical feature would not appear. This particular information unit has three status elements (ACC is engaged, radar is working, and there is a car ahead), three state informational elements (*ACC set speed*, *preceding car speed*, *own speed*), and one additional element – an emergent feature (Δ1 speed between *ACC set speed* and *preceding car* speed). All in all, seven elements are packed into a single unit. It allows the driver in a semi-autonomous vehicle to understand the car's state and status in a quick glance.

Figure 8.2 (b) shows how this integrated unit can be further enhanced to better show the relationships between *ACC set speed*, *preceding car speed*, and *own car speed*. Table 8.1 summarizes the differences between the two integrated structures presented in Fig. 8.1 and Fig. 8.2. We have explicitly highlighted the car's own speed by a step alteration of the arc's shape, making it narrower. This provides two easily identifiable emergent features to the unit: the difference between *set speed* and *own car speed* (Δ2) as well as the difference between *own car speed* and *preceding car speed* (Δ3). On careful observation we also realize that there is another emergent feature here, which is a result of these two deltas; namely their proportional relation −Δ2/Δ3, which provides an indication of how far the *own car* is from closing in on the preceding car. In total, we now have ten informational elements integrated into a single unit (Fig. 8.2 (b)).

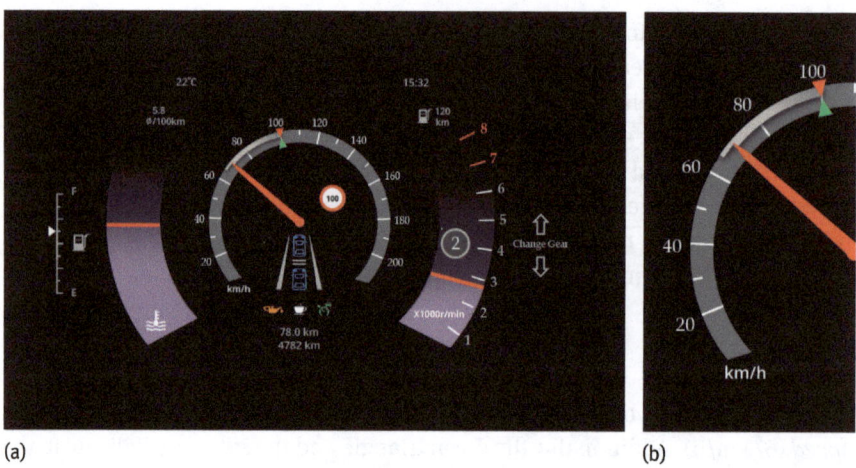

(a)   (b)

Fig. 8.2: (a) An enhanced integrated structure reveals the difference between the "set speed" and the "own car speed" as well as the difference between the "own car speed" and the "preceding car speed". Comparing their relative sizes helps the viewer to quickly understand how far the "own car" is from closing in on the "preceding car". (b) A detailed view of the integrated structure associated with ACC usage.

Table 8.1 compares the two integrative units in Figs. 8.1 and 8.2.

Tab. 8.1: List of informational elements embedded in the two different designs (Fig. 8.2 is a more integrated structure, albeit more complex).

	Figure 8.1	Figure 8.2
1	ACC system is on	ACC system is on
2	Radar is working	Radar is working
3	There is a car ahead	There is a car ahead
4	*ACC set speed* (100 km/h)	*ACC set speed* (100 km/h)
5	*Own car* speed (70 km/h)	*Own car* speed (70 km/h)
6	*Preceding car speed* (67 km/h)	*Preceding car speed* (67 km/h)
7	$\Delta 1$: the difference between *set speed* and *preceding car* speed (100 – 67 = 33 km/h)	$\Delta 1$: the difference between *set speed* and *preceding car* speed (100 – 67 = 33 km/h)
8		$\Delta 2$: the difference between set speed and own car speed
9		$\Delta 3$: the difference between own car speed and preceding car speed
10		$\Delta 2 / \Delta 3$: provides an indication of how far the *own car* is from closing in on the preceding car

In summary, we have shown here that is it possible to segment an integrated structure into a series of basic elements. We have also shown that it is possible to embed more information within an integrated structure by modifying its visual design. Now the question arises, what is the theoretical basis for such integrated structures and how it is possible to create them?

Current approaches of interface design deal primarily with the abstraction of data in order to create separate display representations such as icons and indicator lights. This is often dubbed the "one-sensor, one-indicator" paradigm. There are almost no examples of well integrated units of information in the aviation, automotive and medical domains. One reason is that it takes a considerable ingenuity to create a well-integrated display. While the question of how to design such displays is being referred to in the literature (Wickens & Andre 1990; Bennett, Toms & Woods 1993; Wickens & Carswell 1995), there is little support in terms of tools that can be used in order to create and test them.

The literature provides several theories of integration. According to Bennett and Flach (2011), information integration theories can be classified according to the weight they give to (1) external representation (display elements), (2) information processes (computation processes) or (3) work domain representation. In Fig. 8.3 we present related theories on a Venn diagram with respect to these three dimensions. One can identify a continuum: The Proximity Compatibility Principle

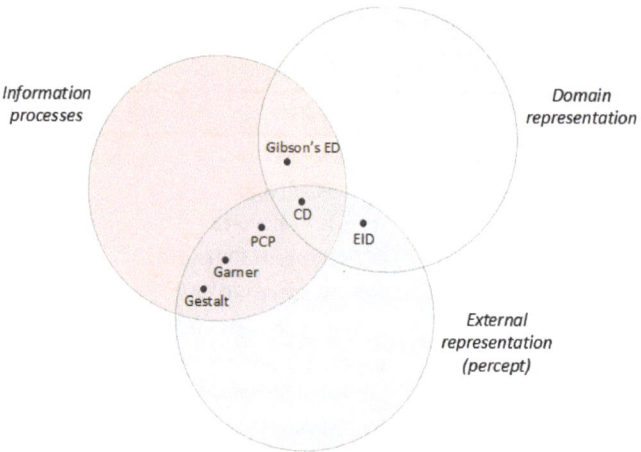

Fig. 8.3: Theoretical map for information integration.

(PCP) ushered in by Wickens and colleagues (Wickens & Andre 1990; Wickens & Carswell 1995) focuses on fairly abstract information processes of external representations, while the domain itself is not an essential dimension of the theoretical framework. We thus place the Proximity Compatibility Principle at the intersection of information processes and the external representation dimension. On the other theoretical extreme is the Ecological Interface Design (EID) theory, ushered in by Rasmussen and Vicente (1989), which assigns considerable importance to domain representation. The theory has its roots in Rasmussen's Abstraction Hierarchy (1985), a functional decomposition of a system into five different levels of abstraction: (i) System Purpose at the top, (ii) Abstract Function, (iii) Generalized Function, (iv) Physical Function, and (v) Physical Form at the very bottom of the hierarchy. Elements at top levels of the model define the goals of the system and the (abstract) functioning of its components and elements at the lower levels describe the physical components of the system and their unique functions. The theory lacks however an account of visual attention and form perception which are nevertheless needed to quantify the integration between the display elements themselves. The Configural Displays (CD) approach advocated by Bennett and Flach (1992) is somewhere in between and draws on all three dimensions. The Configural Displays theory focuses on both high-level constraints in complex systems (relationships among several variables, performance goals, and semantic relationships) as well as low-level data (the values of individual variables and their information representation). According to this theory, the effectiveness of the display depends upon the quality of specific sets of mapping between mutually interacting sets of constraints: the domain (its semantics), the cognitive agent (his/her capabilities and constraints), and the display (the degree to which the representation specifies the affordance of the domain). Both Proximity Compatibility Principle and Configural Displays theory rely on two important theoretical foundations: Gestalt theory (Wertheimer 1923) and the relationships between stimulus dimensions (Garner 1974; Pomerantz & Garner 1973); these two are also presented in the figure, located far from the domain representation. Finally, Gibson's (1979) Ecological Psychology harps on the use of the domain representation and its underlying structure as a foundation for understanding decision-making (where information processing is in a way embedded within the structure of the domain). As such, its implications for display design belongs in the intersection of the domain representation and information processes.

These theories suggest principles for understanding integrative designs, but provide only few guidelines for practical use. In particular, they do not tell us how to analyze integration of information and, more crucially, the necessary steps to form and construct integrated units of information. Here we take a step forward in filling this gap by suggesting a formal approach and a systematic methodology for the analysis and evaluation of integrative displays. Specifically, we aim to quantify the level of integration between informational elements on the instrument display and identify how to improve it.

Wickens and Carswell's Proximity Compatibility Principle provides us with the theoretical framework to quantify relations between informational elements on the display. In its revised version (1995), the Proximity Compatibility Principle depends critically on *perceptual proximity* and *processing proximity*. Perceptual proximity is defined as "*how close together two display elements channels conveying task-related information lie in the user's multidimensional perceptual space*". The second source of proximity, processing proximity, is defined as "*the extent to which two or more sources are used as part of the same task. If these sources must be integrated, they have close processing proximity. If they should be processed independently, their processing proximity is low*". The dimensions of perceptual proximity are the followings: Spatial Proximity, Connections, Source Similarity, Code Homogeneity, Object Integration (Contiguity, Contour and Spatial Integration), and Configuration. High perceptual proximity (e.g. closeness in space, commonality of color, and lines connecting or enclosing two sources) will make comparison and integration easier due to the decrease in *information access cost* (the lower the cost the better the performance). Moreover, this benefit will be enhanced when the search takes place in a noisy and cluttered environment. On the other hand, increases in *information access cost* will disrupt performance on integration tasks.

Car designers use visual means (e.g. color, size, symmetry) to show relationships between informational elements and reflect cause and effects, negative or positive correlations, status classifications, patterns, and trends of different subsystems. Our assumption here is that if the use of visual means can be quantified with respect to the affinity it implies, it should be possible to assess the effectiveness of the design against a hierarchy of clusters using objective methods. Such information can provide a way to quantify the quality of a given design as well as provide insights and opportunities for improved or new designs.

8.2 Method

The method we propose here begins with the abstraction of the vehicle's subsystems data into meaningful task-based representations. The process includes:
1. Listing all informational elements, including those that are currently technically unfeasible.
2. Mapping informational elements into task-based representations that take into account different driving modes and traffic configurations, as well as the driver's cognitive constraints and situation awareness difficulties at times of transfer of control.
3. For each driving configuration, a number of domain experts assess the strengths of the relationships between elements. The result is an affinity matrix that characterizes the strength of the link between pairs of elements.

The methodology is inspired by link analysis (Harper & Harris 1975), a well-known Human Factors technique to evaluate relationships (connections) between nodes, where nodes represent people (in a criminal investigation), machines (shop floor layout design), knobs and dials (in control panel design) and even tasks (in task analysis) in order to minimize or maximize an objective function, for example, reduce the amount of eye movement, increase accessibility, etc. Here we focus our attention on the analysis of information integration in order to help designers determine how a display should be arranged, such that the scanning time between display elements is minimized. The methodology proposed here consists of six steps, as shown in Fig. 8.4.

8.2.1 Listing of informational elements

The content of any given display can be broken down into a set of basic informational elements. For example, consider the car instrument display of Fig. 8.1 (a). The list of informational elements includes items such as speed information (set speed, traffic sign speeds), icons (such as own and preceding car symbols), larger display elements such as the tachometer, water temperature, and fuel level indica-

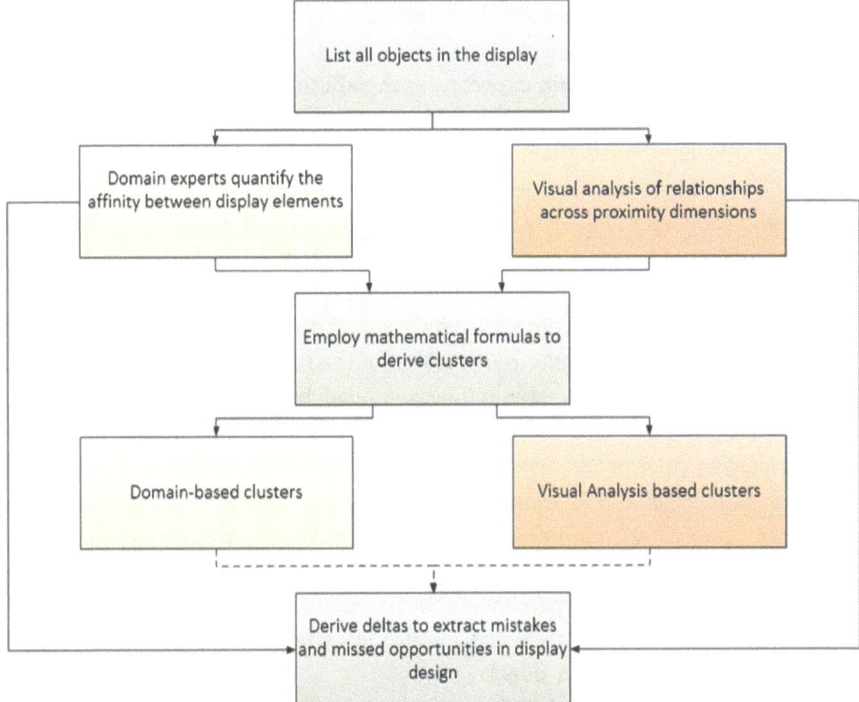

Fig. 8.4: Matching domain-based evaluations with visual-display based evaluations.

tors, as well as alphanumeric information (time, temperature, odometer reading, etc.). The link between the display elements is then analyzed by several domain experts (the light green path) and a design expert (the orange path) as presented in Fig. 8.4.

8.2.2 Domain expert rating

Domain experts are then asked to rate the general affinity between the informational elements. In the example below, we have asked three experts in the automotive field to rate the conceptual links between (all pairwise) informational elements. No display is provided to them (the elements that appear in Fig. 8.1 (a) were provided to the experts in a tabular form, as will be shown later). The affinity values range from 0 (no relationship between display elements) to 1 (extremely strong relationship between display elements). In the analysis described here, the experts' ratings were averaged into a single general affinity matrix.

8.2.3 Measurement of integrative interrelationships

Similarly to the way a domain expert rates the affinity between pairs of informational elements, a design expert quantifies their integrative interrelationship by analyzing a given display along the following dimensions, proposed by Wickens and Carswell's Proximity Compatibility theoretical framework:

1. *Spatial Proximity (Distance)* – the actual (Euclidean) distance between display elements is measured (by pixels, millimeters, or any other unit). There may be different options to define the distance between two objects: as the distance between their centers, as the minimal distance between their borders or as the length of the external segment of the line connecting their centers. Such measures allow to refer to the size and shape of the objects. The results are then inverted (closer elements have higher spatial proximity measure), normalized (to range between 0–1) and optionally manipulated by a function in order to emphasize strong relations and understate weak relations.

2. *Connections (e.g. Contour)* – informational elements that are connected by line segments or enclosed by some visual contour (e.g. own car and preceding car symbols inside the speedometer in Fig. 8.1 (a) are ranked high on this *connections* dimension because they appear as a single unit due to the inclusion of the lane lines on both sides).

3. *Source Similarity (e.g. Color)* – analog elements that share the same color, orientation and perspective as well as informational elements sharing the same font (e.g. the tachometer and engine temperature gauges have the same color, shape, and orientation and thus are ranked high on source similarity).

4. *Code Homogeneity (e.g. Digital/Analog)* – objects sharing a visual property such as analog gauge and digital format (e.g. the coffee mug and oil icons inside the speedometer are both icons that share code homogeneity).

5. *Object Integration* – display elements that appear to the user to be part of a more structured and rich object because of their contiguity (e.g. the green set speed marker and the red speed sign marker are both triangles that, when aligned, form a single integrated object).

6. *Configuration* – homogenous elements that generate a new pattern (e.g. on the analog speedometer, the silver arc between the set speed and current speed is an emerging pattern).

For each dimension, a design expert quantifies the interrelationship between all pairwise elements. A value of one is assigned to integrated pairs and zero otherwise. The affinity is calculated as a weighted sum of the considered dimensions (equation (8.1)). Our weights for the distance, contour, color, digital/analog, symmetry and integration reflect the relative importance of each dimension to the visual integration measure. These were $w_{dist} = 3$, $w_{cont} = 1$, $w_{col} = 2$, $w_{dig/ang} = 1$, $w_{sym} = 2$, $w_{int} = 2$, respectively (these weights can be considered as a possible input of the method, and are subject to further research).

$$affinity = w_{dist} \cdot distance + w_{cont} \cdot contour + w_{col} \cdot color$$
$$+ w_{dig/ang} \cdot dig/ang + w_{sym} \cdot symmetry + w_{int} \cdot integration \tag{8.1}$$

8.2.4 Clustering algorithm

A clustering algorithm is used to evaluate the connectivity between elements in a given system in order to identify salient subsets that share a similar or related functionality (Albert et al. 2002; Newman 2006; Strogatz 2001). The clustering algorithm first constructs a graph of nodes and edges where the nodes refer to the different informational elements and edges are assigned weights that represent the connectivity strength between any two elements. The algorithm ranks subsets of the elements in a pyramid-like structure.

The input for the algorithm is a set of K symmetric adjacency matrices that describe the pairwise relations between the n elements of the system with respect to K different features (e.g. different integrative dimensions or experts in our case) together with the general weight of each feature. A global matrix A is then calculated as a function of the weighted sum of adjacency matrices:

$$A_{i,j} = f \left(\frac{\sum_{k=1}^{K} w_k C_{i,j}^k}{\sum_{k=1}^{K} w_k} \right), \tag{8.2}$$

where $0 \leq C_{i,j}^k \leq 1$ describes the connectivity between the ith and j^{th} elements with respect to the kth feature, and w_k is the feature weight. The function f is a logistic function of a sigmoidal shape:

$$f(t) = \frac{1}{1 + e^{-B(t-M)}}, \tag{8.3}$$

where $B(= 27)$ and $M(= 2/3)$ are positive constants that define the rate and the position of the sigmoidal slope, respectively (Fig. 8.5). The logistic function preserves the higher values whereas lower values are filtered such that salient clusters will emerge due to their stronger connections.

We then consider our data as a graph $G(V, E)$, whose nodes V are the n informational elements and the weights of the edges E are defined by the symmetric $A_{n \times n}$ matrix. Two objective functions are applied to every subset $S \subseteq V$. The clique function measures the connectivity between the nodes (Fig. 8.6), by normalizing the sum of the weights of the edges within S:

$$clq\,(S) = \frac{\sum_{i,j \in S} A_{i,j}}{n_S(n_S - 1)} \in [0, 1]. \tag{8.4}$$

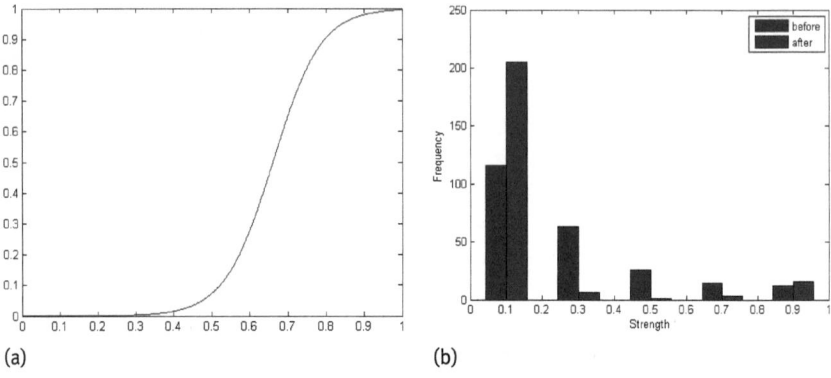

(a)   (b)

Fig. 8.5: The logistic function (left) is a sigmoidal function that filters values by shifting values lower than the position of its slope towards the minimum of the range of its domain (right).

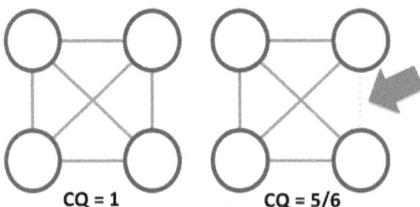

CQ = 1   CQ = 5/6

Fig. 8.6: Based on the clique function, a subset of four elements takes a maximal clique value of 1 when it is fully connected (left) and lower values when some edges are missing (right).

Large clusters with a relative high clique measure (*clq*) are found by utilizing the cluster size:

$$f_{clq}(S) = n_s^a \cdot clq\,(S)^{n_s^b}, \tag{8.5}$$

where $a(=1)$ and $b(=0.4)$ are positive constants.

The modularity function measures, for every external node $v_e \in V \setminus S$, the similarity between the weights of its edges to the nodes in S:

$$f_{mod}(S, v_e) = \mu - 3\sigma, \tag{8.6}$$

where μ and σ are the mean and standard deviation of the weights of these edges. The coefficients in the equation were chosen such that the function is a plane defined by the (μ, σ, f) triplets: $(0, 0, 0)$, $(1, 0, 1)$, and $(0.5, 0.5, -1)$. Figure 8.7 presents the modularity score of alternative four-node configurations.

A global function F combines the *clique* score with the summation of the *modularity* score over the subset's external nodes:

$$F(S) = f_{clq}(S) + c \sum_{v_e \in V \setminus S} f_{mod}(S, v_e), \tag{8.7}$$

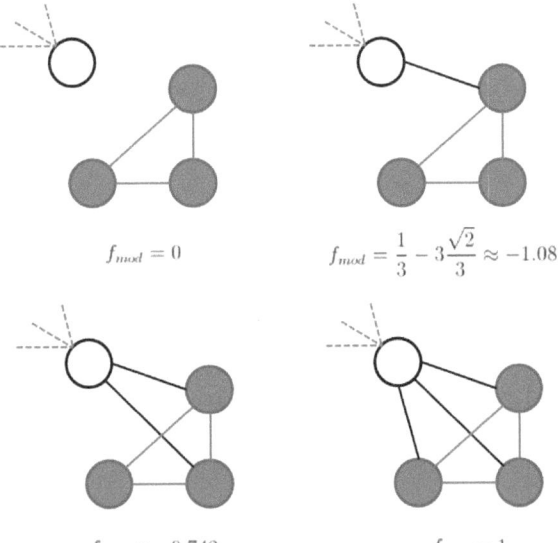

$$f_{mod} = 0$$

$$f_{mod} = \frac{1}{3} - 3\frac{\sqrt{2}}{3} \approx -1.08$$

$$f_{mod} \approx -0.748$$

$$f_{mod} = 1$$

Fig. 8.7: Modularity score.

where $c(= 1/3)$ is a constant that reflects the ratio between the weights of the measures. This global F function evaluates how well a subset can function as a cluster, such that its nodes are both connected to each other and connected to the rest of the graph with similar patterns of weights. Finally, the subsets are sorted by their global F score and a hierarchical pyramid structure is constructed, such that a cluster with a larger score is placed higher.

8.2.5 Comparison of the two hierarchical structures

The clustering algorithm is applied on the two affinity matrices (the domain expert matrix and the design expert matrix) resulting in two hierarchical structures. The domain experts' hierarchical structure reveals the relations between informational elements independently of the interface on which they are being displayed. The design experts' hierarchical structure represents the underlying design of the actual interface.

8.2.6 Comparisons between the domain expert and the design expert analyses

Quantitatively, the absolute difference between the cells of the two affinity matrices is then used to identify disparities. When the difference between each pair is relatively small, we can say that the domain experts' rating agrees with the design expert's rating. When the difference is relatively large, it indicates a certain mismatch between the experts' ratings and the display measurement. An inspection of the differences between the two cluster hierarchies may help to understand the source of any such mismatch within the overall design context.

8.3 Analysis of an instrument display

We now illustrate this methodology by analyzing the instrument display of Fig. 8.1 (a).

We begin the analysis with **Step 1** by listing all basic informational elements of the display. Table 8.2 is a breakdown of the instrument display of Fig. 8.1 into its relevant elements of information (see Tab. 8.2).

In **Step 2** three experienced domain experts evaluated and rated the conceptual relationship between each pair of elements listed in Tab. 8.2. The ratings provided by the three experts were averaged for purposes of this analysis (other summary statistics could also be considered). Figure 8.8 presents their averaged values.

Tab. 8.2: List of informational elements used in the analysis.

1	Own speed	Speedometer
2	Ext speed	External speed, or speed limit
3	Own car	Own car symbol in the ACC pyramid
4	ACC set speed	ACC state
5	Alerts	All sorts of alerts
6	ACC on/off	ACC status
7	Fuel Consump.	Average fuel consumption
8	Fuel left	For a number of km
9	Fuel level	Tank level
10	Preceding	Preceding car symbol in the ACC pyramid
11	Gap	Gap between own car and preceding car
12	Lanes	lanes in the ACC pyramid
13	Road info	Traffic signs or external speed limit
14	Engine temp	Engine temperature
15	Ext temp	Outside temperature
16	Total km	Total km of the car
17	Ride km	km in this ride
18	Time	Time – HH:MM
19	Gear	Gear state
20	Tach.	Tachometer
21	Change gear	Change gear indicator

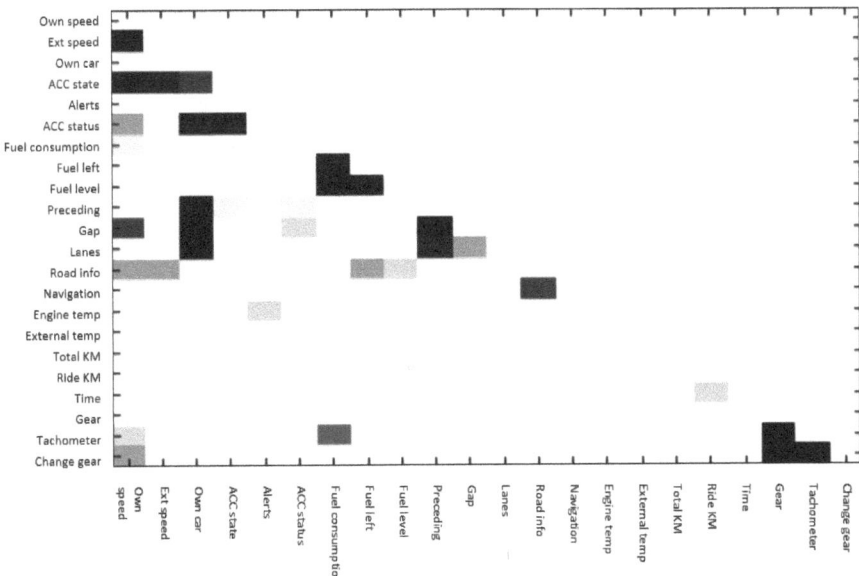

Fig. 8.8: Averaged values derived from the domain experts' analysis of the generic cluster display – the darker the color, the stronger the affinity link.

Step 3 involved the measurement of integration between all the informational elements as being reflected by the six integrative interrelations dimensions (according to equation (8.1)). Figure 8.9 presents the values derived from the design analysis of the generic cluster display.

Applying the clustering algorithm (**Step 4**) separately on the domain expert ratings and the visual-analysis ratings yields two hierarchical clusters. Figure 8.10 shows the hierarchy of clusters achieved from the domain experts' ratings and Fig. 8.11 shows the hierarchy of clusters achieved from the visual-analysis ratings. The clusters are labeled sequentially according to the strength of the connections (edges) within each cluster.

Inspection of the domain experts' cluster in Fig. 8.10 suggests that several units of information are candidates for integration. With respect to automated driving features (adaptive cruise control and lane keeping), the data suggests that it may be advantageous to lump together the informational elements representing the preceding car, own car, gap (time-distance), and traffic lanes (*group 3*) into an integrated unit. Furthermore, the state information such as own speed, external speed (traffic sign element), and ACC set speed can also be presented in a way that helps the driver view their interrelationship as connected elements (*group 4-5*).

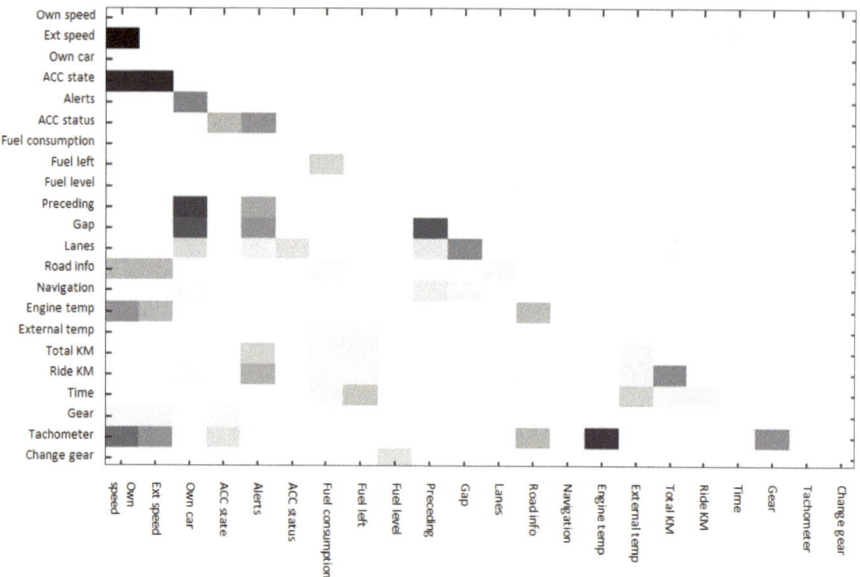

Fig. 8.9: Averaged values derived from the design analysis of the generic cluster display – the darker the color, the stronger the affinity link.

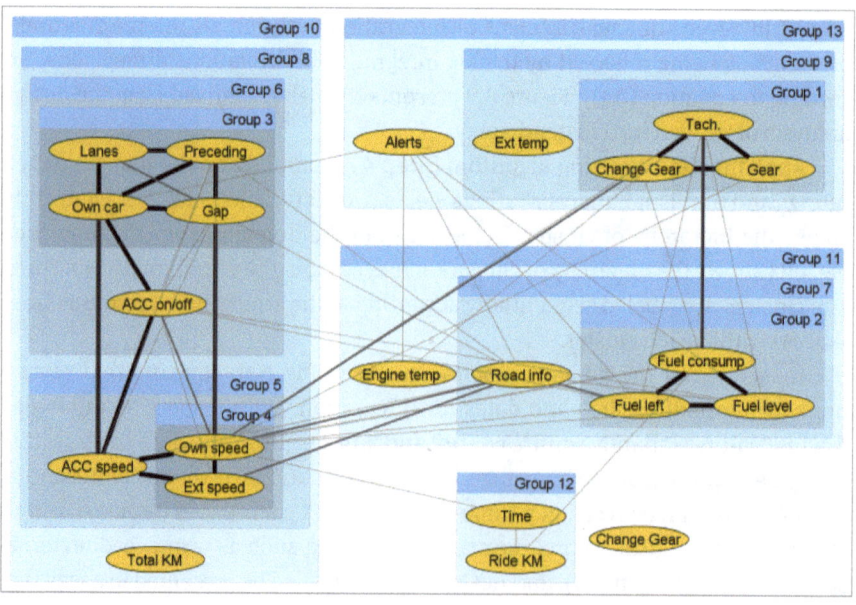

Fig. 8.10: Hierarchical clusters derived from the domain experts' ratings.

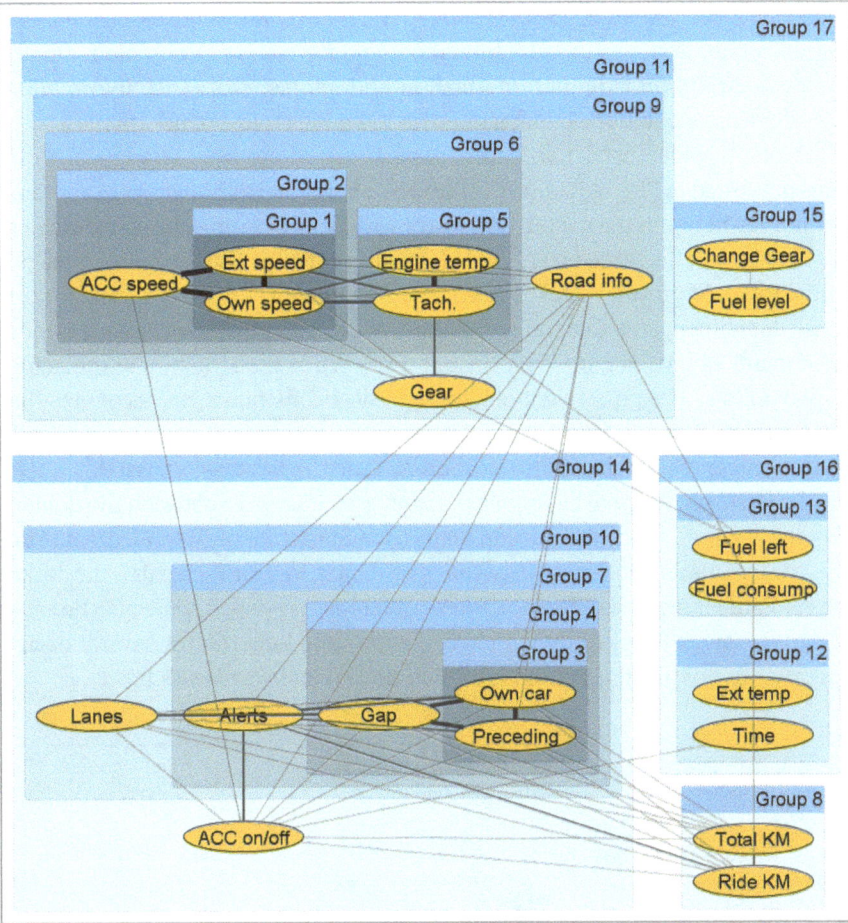

Fig. 8.11: Hierarchical clusters derived from the analysis of the display.

This hierarchical structure also presents the ties between pairs of informational elements (and their strength – denoted by the width and shade of the lines). Analysis of the edges can provide insights into how the above two units (*group 3* and *group 4-5*) could/should be connected. For example, the ACC on/off indicator is related to both groups and could serve as a visual connector.

Figure 8.11 presents the hierarchical cluster that emerged from the integration analysis along all six integrative interrelations dimensions described earlier. The resulting hierarchical structure indicates that the design of the instrument display capitalizes on three main groups of information. The first consists of *groups 3-4-7-10-14* (associated with ACC and lane information, as well as several alerts). The second includes *groups 1-2-5-6-9-11-17*, which are associated with speeds, tachometer, gear, road information, and navigation information. Finally, *groups 8-12-13-16* provide fuel information and odometer/distance information. The edges show additional connections between the groups of information.

A side-by-side, inspection-based, comparison of the two hierarchical structures presented in Figs. 8.10 and 8.11 (**Step 5**), shows that although there are some similarities between the overall structures with respect to specific elements (e.g. ACC and lanes), the two clusters are in fact quite different. This suggests some potential for an improved visual display, which can better support the observations of the domain experts. Generally speaking, the hierarchical structure created by the domain experts is more concise and tight. While the structure of the actual display follows design conventions and physical constraints, we recognize that some opportunities for integration of informational elements that are practically and/or conceptually related into a coherent structure have been missed.

The last step (**Step 6**) examines the numeric differences between the domain experts' ratings and the integration analysis in the actual display. Figure 8.12 details the (absolute) differences between the two matrices for each pair – the darker the color, the larger the difference. These differences revealed some questionable design decisions (marked by red circles) and, more importantly, several design opportunities of the integrative kind (blue circles) that we discuss now:

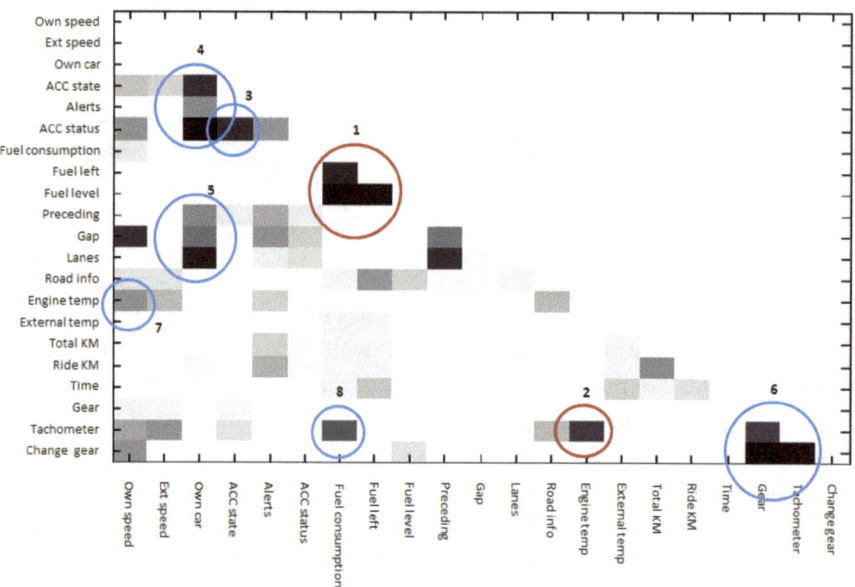

Fig. 8.12: Matrix subtraction (absolute values).

1. *Fuel related information:* The fuel level in the fuel tank appears on the left using a conventional, analog representation. The average fuel consumption appears in the upper-left area using alphanumeric representations. On the other hand, the distance to empty fuel tank, calculated using these two parameters, appears in the upper-right area, which is distant from both (i.e. low proximity manipulations). When the driver wants to know how the fuel remaining in tank affects the "distance left to drive", he needs to scan two distant locations and mentally merge the analog data with the alphanumeric information. Similarly, the assessment concerning the average fuel consumption on the drop-rate of fuel in the tank requires complex mental inference, although with less effort given the smaller distance between the two elements. A better design strategy might be to arrange all fuel related elements in a single configural object to match the semantics of the task, with the alphanumeric information integrated within the object.

2. *Tachometer – Engine Temperature:* Several proximity manipulations were made to make these two elements highly related. As the hierarchical cluster of Fig. 8.11 shows, these elements form a clique (*group 5*), as a result of employing color, shape symmetry, analog indicators, and a common manipulation of object integration. The question is whether these efforts provide any useful relations to the driver?

3. *ACC Status (ACC Set Speed) – ACC State (on/off):* Other than the common color, there is no similarity between these two operationally related elements. They are also situated at a distance from each another (low spatial proximity).

4. *Own Car – ACC Status – ACC State:* There is no visual relationship between the car symbol in the center of the speedometer and the green icons of the ACC. (One could perhaps create some radar lines cueing the ACC Status/State color to explicitly relate between the three elements).

5. *Own Car – Preceding Car – Traffic Lanes:* Designers might consider creating some display icons (e.g. chevrons) to emphasize the car's relative position between the lanes and increase the perceived proximity of the two informational elements.

6. *Tachometer – Gear – Change Gear:* The tachometer in the display incorporates the current gear information, which is a good form of integration. While these three elements are currently placed in fairly close spatial proximity, assigning the shift gear indication (up/down arrows) sign with the gear state could make this triplet a highly integrative unit.

7. *Own Speed – Engine Temperature:* The spatial proximity between the elements and the red needle make them appear more related than they actually are.

8. *Tachometer – Fuel Consumption:* There is a strong dynamic and physical interrelationship between high engine revolution (observed on the tachometer) and increased fuel consumption. This provides an opportunity to link the two and reveal the interrelationships to the ("eco"-minded) driver.

8.4 Conclusion

The ever-growing number of information and automation features in modern cars and the complex relationships between them, have turned the vehicle into a sophisticated system with many informational elements that need to be displayed and monitored. Concurrent advances in display technology and the introduction of a flat LCD screen have freed designers to organize cluster information in any shape, color, or suitable arrangement. This freedom can easily lead to a hodgepodge of display organizations unless there is a solid theoretical approach for information organization that allows for abstraction, integration, and configuration of data and information for effective interface design.

This chapter is a step towards such a theoretical approach by providing a framework for analysis of visual interfaces and a formal methodology for its computation. The approach presented here quantifies the relationships between display elements by employing link analysis and graph theory on two sets of data: (i) an affinity matrix that quantifies the advisable strengths of conceptual links between informational elements as determined by domain experts, and (ii) an affinity matrix that quantifies the actual values derived from the display under consideration. Large differences between pairwise elements of matrices (i) and (ii) indicate a mismatch between the experts' ratings and the display measurement and hence reflect suboptimal design solutions. In this chapter, we demonstrated this approach via an analysis of a generic cluster display. We believe that the approach and method will enable design practitioners to evaluate existing displays and identify opportunities for integration of information. This could lead to the ability to explicitly point out design mistakes and identify missed opportunities. Mistakes could then be amended and the improved understanding of the missed opportunities may lead to new, and better integrated, design structures and concepts.

In order to be used optimally, the proximity manipulations presented in this chapter must derive from a full understanding of task constraints (the nature of the work to be done), and display constraints (the spatial and temporal characteristics of the visual form). For example, if the scale used for representing the car speed is incompatible with the user's mental model of car speed, or if the icons of the display inappropriately represent the selected domain, then a correct ar-

rangement of a display in terms of objects' location, color, shape, or object configuration will not be sufficient to make a good interface. The compatibility between
physical function and form, as entailed by the Ecological Interface Design theory (Rassmusen & Vicente 1989; Vicente 1999), must be well understood to verify
that the complex array of relations between elements (e.g. cause and effect, direct
relationships, inverse relationships, etc.) is being represented successfully.

Understanding the compatibility between physical function and form is also
needed in order to successfully represent the relative importance of each informational element and to choose the right visual representation to reflect this correctly. Formulating an additional equation, that will successfully calculate the individual importance of a display element as a weighted sum of its visual attributes,
can be used not only to account for the general appearance of the elements but
also to explain the need for emphasized appearance in urgent situations. Still, the
solution should be verified to ensure that the optimal affinity between information
elements is being preserved in spite of such temporary changes.

The method can be further extended to support the dynamic range that may
be associated with advanced interfaces. By taking into account the display transitions required during dynamic driving scenarios, we can calculate a series of
affinity matrices that correspond to sequential time points in a scenario. This can
be used by a design expert to verify that the dynamic scenario is being displayed
correctly; that in every given moment the observations made by the domain expert
are well preserved at the display level and that the display transitions are smooth
and well made.

Finally, in bringing the discussion on the evaluation of displays and creation
of well integrated structure, we would like to make the point that capturing relations between information elements is only the foundation of integrated and well
configured displays. Creating good visual layouts as well as integrative designs is
a difficult task which requires a deep functional understanding of the elements
that are integrated and also a certain emotional sensitivity to associate elements
in a coherent way (Jung 1955/1972). Nevertheless, we believe that the integration
of informational elements and their configuration to create a structured whole is
the key to building the information-intensive interfaces of the future.

8.4.1 Extension of the approach to sound- and haptic-interfaces

Sound and haptics have the potential to direct the driver's attention to car internal and external events in subtle or explicit manner, depending on the semantics
of the events and the driver's characteristics. Automotive companies have been
using chimes (as an acoustic interface) and haptics (as a vibrotactile interface) to

warn the driver, remind, notify, or provide feedback for certain events or situations, manipulating their characteristics to express varying levels of urgency. Until fairly recently, vehicle chimes were generated by the speaker-based system "on the fly" from predefined, fixed sound frequencies and amplitude profiles. These dimensions, as well as the sound cadence period, the number of repetitions and the sound duty cycle of each tone, were used to reflect different event categories (reminders, maintenance alerts and active safety alerts) with different urgency values (Nees & Walker 2011). The array of the vehicle loudspeakers has been used to convey spatially-localized sounds for an increased situation awareness and improved response times, contributing to the semantics of the chimes (Ho & Spence 2005). Haptic interfaces provide alternative forms for conveying spatial information to car drivers and are currently receiving a great deal of both empirical and commercial interest (Ho & Spence 2008). Tactile warning signals were found to be more effective than equivalent warning signals presented by other modalities (Meng & Spence 2015; Fitch et al. 2007; Prewett et al. 2012). It has been shown, for example, that compared with sound-based collision warnings, tactile warnings are effective during a phone conversation (Mohebbi et al. 2009). Haptic interfaces may enhance the driver experience by delivering a great deal of information rapidly and simultaneously (Lee et al. 2004). They may be very effective especially while operating cars with automated driving features that are most likely to encourage secondary driving tasks and even non-driving activities. Specifically, vibrotactile displays were found to be effective and intuitive in terms of presenting directional information (Tan et al. 2003) and navigation directions (Van Erp & Van Veen 2001), localizing alerts (Terrence et al. 2005), and improving the awareness of the driver to the surrounding traffic (Telpaz et al. 2015).

Sound and haptics have several attributes in common: speed, frequency, intensity, tempo, regularity, fluctuation strength, and number of repetitions. In addition, sound has some unique modality attributes, such as harmonicity, roughness, and tonality. Understanding the relative importance of these spectral and temporal psychoacoustic features in creating a sense of urgency (Edworthy et al. 1991), or in forming a semantic characterization of the event (Kleiss et al. 2012) can serve as a basis for the generalizations of the proximity classification and mathematical tools, described above, to these additional modalities. If sound and haptics parameters can be quantified, it may be possible to analyze each of their set of attributes in the same manner as we have previously done for the visual domain and verify that they match the semantics of the event and the driver's requirements. But there is something more here: since they share many attributes, it may be possible to combine the two modalities and consider their cross-modal effect (for a detailed review of the effect of spatial congruency on audio-tactile multimodal cues, see Ho & Spence 2008, Ch. 7), which brings us to the last point in this discussion – the use of the approach presented above to quantify displays with several different modalities.

8.4.2 Multimodal presentation

We believe that it is possible to extend the method detailed above to the problem of multimodal presentation configurations, supporting a comprehensive analysis that considers sound, speech and haptics together with the visual display elements. This would require both the proximity classifications and the mathematical tools to refer to the additional modalities, when considered individually and as multimodal compounds.

Looking at the combined use of visual, sound and haptics modalities in the automotive industry, reveals fairly basic multimodal designs: traditionally, either chimes or haptics were used: (i) to guide the user's attention to the car's visual displays that in turn, provided further information in the form of tell-tales, textual messages, change of color etc., and (ii) to provide spatial information regarding the location of surrounding objects, where applied. Nevertheless, recent advances in vehicles' electric architecture enable manufacturers to shift from chimes to pre-recorded sound files, providing designers opportunities to create sophisticated auditory alerts and notifications. Furthermore, advances in sensing technology provide longer preparation durations before an event takes place, thereby opening the door for a dynamic selection of modalities as well as for gradual increment in the sense of urgency that an individual modality, or multimodal compounds can produce depending on the current context.

As an example, a collision warning display may consist of multiple visual elements that are being distributed around a vehicle image, referring to the entire surrounding of the vehicle. The affinity between these elements is clearly high and therefore a similar visual design is expected for all. However, in some cases we can expect some of the informational elements to get a higher importance rating in order to draw attention to a possible danger from a specific direction (e.g. when the vehicle is entering a road segment in which other vehicles are usually merging from the right side of the vehicle). Furthermore, in case of an actual danger of collision, we clearly expect the appropriate element to change its features to signal an alert. The diversity of tools associated with the different modalities allows us to develop a principal approach regarding their combined usage. Considering the above example, visual similarity can be used to present the elements in their default state, whereas changing just one visual characteristic will direct the user attention to a required location. In addition, a localized, mild vibrotactile signal may provide a redundant subtle cue to orient the driver's visual attention to the appropriate location. In case of emergency, the designer may change the visual characteristics of the display element dramatically, amplify the characteristics of the vibrotactile feedback, and also add a spatially-localized urgent sound to signal an alert. Overall, these various tools give the ability to follow constraints of the interrelations between elements during dynamic events without compromising the importance of each element separately; they allow us to provide a holistic interface that maximizes the utility of each modality, depending on the situation and the driver state.

In the context of the approach suggested here, one can adjust the affinity equation to include all the attributes of the three modalities. This will enable multimodal design experts to assess the effectiveness of their designs against a hierarchy of clusters in a similar manner to the comparison demonstrated between the visual integration hierarchical structure and the domain experts' hierarchical structure. Similar principles will also form design opportunities to unify acoustic, haptic and visual information elements into coherent structures using judicious multimodal strategies, and to select optimal combinations for different users.

Abbreviations

ACC   Adaptive Cruise Control
CD   Configural Displays
PCP   The Proximity Compatibility Principle
EID   Ecological Interface Design

References

Alexander, C 2002, *The Phenomenon of Life*, The Center for Env. Structure, Berkley.
Albert, R, Barabási, AL 2002, Statistical mechanics of complex networks, *Reviews of Modern Physics*, vol. 74, no. 1, pp. 47–97.
Barshi, I, Degani, A, Iverson, D & Lu, PJ 2012, Using medieval architecture as inspiration for display design, *Proceedings of the Human Factors and Ergonomics Society Annual Meeting*, vol. 56, pp. 1799–1803.
Bennett, K & Flach, J 1992, Graphical displays: Implications for divided attention, focused attention and problem solving, *Human Factors*, vol. 34, no. 5, pp. 513–533.
Bennett, K & Flach, J 2011, *Display and Interface Design: Subtle Science, Exact Art*, CRC Press, Boca Raton, FL.
Bennett, KB, Toms, ML & Woods, DD 1993, Emergent features and configural elements: Designing more effective configural displays, *Human Factors*, vol. 35, pp. 71–97.
Broy, M, Krüger, I, Pretschner, A & Salzmann, C 2007, Engineering Automotive Software, *Proceedings of the IEEE*, vol. 95, no. 2, pp. 356–373.

Charette, R 2009, This Car Runs on Code, *IEEE Spectrum*, February Issue, http://spectrum.ieee.org/green-tech/advanced-cars/this-car-runs-on-code.

Degani, A 2013, A Tale of Two Maps: Analysis of the London Underground "Diagram", *Ergonomics in Design: The Quarterly of Human Factors Applications*, vol. 21, pp. 7–16.

Degani, A, Jorgensen, C, Iverson, D, Shafto, M & Olson, L 2009, *On Organization of Information: Approach and Early Work*, NASA TM 2009–215368, NASA-Ames Research Center, Moffett Field, CA.

Dingus, TA & Hulse, MC 1993, Some human factors design issues and recommendations for automobile navigation information systems, *Transportation Research*, Part C, IC, pp. 119–131.

Edworthy, J, Loxely, S & Dennis, I 1991, Improving auditory warning design: Relationship between warning sound parameters and perceived urgency, *Human Factors*, vol. 33, no. 2, pp. 205–231.

Fitch, GM, Kiefer, RJ, Hankey, JM & Kleiner, BM 2007, Toward developing an approach for alerting drivers to the direction of a crash threat, *Human Factors: The Journal of the Human Factors and Ergonomics Society*, vol. 49, no. 4, pp. 710–720.

Garner, WR 1974, *The Processing of Information and Structure,* Lawrence Erlbaum Associates, Potomac, MD.

Gibson, JJ 1979, *The Ecological Approach to Visual Perception*, Houghton-Mifflin, Boston, MA.

Harper, W & Harris, D 1975, The Analysis of Criminal Intelligence, *Proceeding of the Human Factors and Ergonomics Society Annual Meeting*, vol. 19, no. 2, pp. 232–238.

Ho, C & Spence, C 2005, Assessing the effectiveness of various auditory cues in capturing a driver's visual attention, *Journal of Experimental Psychology: Applied*, vol. 11, pp. 157–174.

Ho, C & Spence, C 2008, *The multisensory driver: Implications for ergonomic car interface design*, Ashgate, Aldershot.

Jung, CG 1955/1972, *Mandala Symbolism,* Princeton University Press, Princeton.

Kleiss, JA, Georgiev, EM & Robinson, SW 2012, "Semantic Profiling: A Method for Relating Auditory Device Signals and Medical Messages", in *Symposium on Human Factors and Ergonomics in Health Care: Bridging the Gap*, Human Factors Society, Baltimore, Maryland, USA. pp. 202–206.

Labuhn, PA & Chundrlik, WJ 1995, *Adaptive Cruise Control, General Motors*, Patent No. 5454442, Filing Date: Nov 1, 1993, Issue date: Oct 3, 1995.

Lee, JD, Hoffman, JD & Hayes, E 2004, Collision warning design to mitigate driver distraction, in *Proceedings of the SIGCHI Conference on Human Factors in Computing Systems*, pp. 65–72, ACM, New York.

Lu, PJ & Steinhardt, PJ 2007, Decagonal and Quasi-Crystalline Tilings in Medieval Islamic Architecture, *Science*, vol. 315, p. 1106.

Meng, F & Spence, C 2015, Tactile warning signals for in-vehicle systems, *Accident Analysis & Prevention*, vol. 75, pp. 333–346.

Mohebbi, R, Gray, R & Tan, HZ 2009, Driver reaction time to tactile and auditory rear-end collision warnings while talking on a cell phone, *Human Factors: The Journal of the Human Factors and Ergonomics Society*, vol. 51, no. 1, pp. 102–110.

Nees, MA & Walker, BN 2011, Auditory Displays for In-Vehicle Technologies, *Reviews of Human Factors and Ergonomics*, vol. 7, no. 1, pp. 58–99.

Newman, M 2006, Modularity and community structure in networks, *Proceedings of the National Academy of Sciences of the United States of America*, vol. 103, no. 23, pp. 8577–8582.

Pomerantz, JR & Garner, WR 1973, Stimulus configuration in selective attention tasks, *Perception and Psychophysics*, vol. 14, pp. 565–569.

Prewett, MS, Elliott, LR, Walvoord, AG & Coovert, MD 2012, A meta-analysis of vibrotactile and visual information displays for improving task performance, *Systems, Man, and Cybernetics, Part C: Applications and Reviews, IEEE Transactions on*, vol. 42, no. 1, pp. 123–132.

Rasmussen, J & Vicente, K 1989, Coping with human errors through system design: Implications for ecological interface design, *International Journal of Man-Machine Studies*, vol. 31, pp. 517–534.

Rasmussen, J 1985, The role of hierarchical knowledge representation in decision making and system management, *IEEE Transactions on Systems, Man and Cybernetics*, vol. 15, pp. 234–243.

Shmueli, Y, Degani, A, Zelman, A, Asherov, R, Zande, D, Weiss, J & Bernard, A 2013, Toward a formal approach for information integration: Evaluation of an automotive instrument cluster, *Proceedings of the 57th Annual Meeting of the Human Factors and Ergonomics Society*, Human Factors Society, San Diego, CA.

Strogatz, S 2001, Exploring complex networks, *Nature*, vol. 410, p. 6825.

Tan, HZ, Gray, R, Young, JJ & Traylor, R 2003, A haptic back display for attentional and directional cueing, *Haptics-e*, vol. 3, no. 1, pp. 1–20.

Telpaz, A, Rhindress, B, Zelman, I & Tsimhoni, O 2015, Haptic seat for automated driving: preparing the driver to take control effectively, in Burnett, GE, Gabbard, JL, Green, P & Osswald, S, eds, *Proceedings of the 7th International Conference on Automotive User Interfaces and Interactive Vehicular Applications*, 'AutomotiveUI', ACM, pp. 23–30.

Terrence, PI, Brill, JC & Gilson, RD 2005, Body orientation and the perception of spatial auditory and tactile cues, in *Proceedings of the Human Factors and Ergonomics Society Annual Meeting*, Sage Publication, Orlando, Florida, pp. 1663–1667.

Van Erp, JB & Van Veen, HAHC 2001, Vibro-tactile information presentation in automobiles, in Berber, C, Faint, M, Wall, S & Wing, AM, eds, *Proceedings of Eurohaptics*, University of Birmingham, Birmingham, pp. 99–104.

Vicente, KJ 1999, *Cognitive work analysis: Toward safe, productive, and healthy computer based work*, Erlbaum, Mahwah, NJ.

Wertheimer, M 1923, *Laws of organization in perceptual forms*. Published in "A source book of Gestalt psychology", pp. 71–88, Routledge & Kegan Paul, London.

Wertheimer, M 1923, Laws of organization in perceptual forms. First published as Untersuchungen zur Lehre von der Gestalt II, in *Psycologische Forschung*, vol. 4, pp. 301–350. Translation published in Ellis, W 1938, *A source book of Gestalt psychology*, pp. 71–88, Routledge & Kegan Paul, London. [available at http://psy.ed.asu.edu/~classics/Wertheimer/Forms/forms.htm]

Wickens, C & Andre, A 1990, Proximity compatibility and information display: Effects of color, space, and abjectness on information integration, *Human Factors*, vol. 32, pp. 61–78.

Wickens, C & Carswell, C 1995, The proximity compatibility principle: Its psychological foundations and its relevance to display, *Human Factors*, vol. 37, no. 3, pp. 473–494.

Michael Heymann and Asaf Degani

9 Classification and organization of information

The case of the head up display

Abstract: The automotive industry has been undergoing a massive, rapid change involving new and emerging technologies, including automatic driving aids and safety enhancement features. One clear outcome of this change is the need to provide drivers with sound organization of information to allow them to handle the growing in-vehicle interface complexity safely and effectively. This chapter address the issue of information organization in automotive cockpits from a formal perspective. It begins by discussing the informational elements available for display and the advantages of applying a function (additive, weighted, or otherwise) to generate a priority ranking scale. We then show how this categorization could be used to evaluate the allocation of information elements to displays (cluster, center stack, etc.) in the car. In the case of a head up display (HUD), once the list of information elements designated for display is agreed upon, the question of its organization given the *benefit* of such a display (information in the main field of view, ease of access) and *cost* (distance from line of sight, number of steps, time to completion) is discussed. We suggest several guiding principles for head up display organization and review several existing designs.

9.1 Introduction

In the past two decades automobiles and the automotive industry as a whole have been undergoing a massive, rapid change involving the implementation of new and emerging technologies. These range from changes in vehicle power plants and propulsion systems, vehicle stabilization and safety control to technologies aimed at reducing the driver's workload, increasing passenger safety, comfort, entertainment and communication capabilities. Some of these changes can be attributed to the extensive use of computing power in all vehicle systems, whereas others have been driven by the emergence of advanced and affordable sensing devices associated with powerful information processing technologies.

These unprecedented advances afforded multitude of features and enhancements such as engine control systems to achieve efficiency and performance, computer controlled braking systems, as well as vehicle stability and traction control, active power steering and automated braking, automated parking assistance to enhance the driving and handling experience, infotainment systems to allow access to digital content and Internet, and navigation systems and location services to maintain contact with the outside world (Leen & Hefferman 2002). All these technologies have had an immediate, profound impact on the driver's operating environment and on the design of the cockpit. In particular, they place increased demands on the driver (and to some extent also the passengers) to understand, monitor, and interact with these systems (Broy et al. 2007). Moreover, there has been a major thrust to equip vehicles with a variety of automated driving aids (Norman 2007), sometimes referred to as Advanced Driver Assistance Systems (or ADAS). These include standard cruise control to maintain constant speed (which has been in existence for several decades), adaptive cruise control and full speed range ACC to maintain a safe distance from preceding vehicles that was introduced more recently, and automatic lane centering control that is in its early phases of commercial deployment. Along with these capabilities there are a variety of driver alerting and warning devices that are aimed at increasing driver alertness, vehicle safety, and overall road safety (Minoiu-Enache et al. 2009; Raphael et al. 2011). In recent years, increasing efforts have been made to develop technologies that will lead to automated and ultimately to autonomous driving.

Concurrently, rising traffic congestion and the resulting time delays force drivers to spend increasing amounts of time in their vehicles. As a result, there has been an ongoing shift in the way drivers and especially commuters view the time spent in their cars. People tend to consider this time as an opportunity to communicate with colleagues, friends, and family (Richtel 2011), work, or to enjoy the benefits of advanced infotainment systems. However, all these activities can distract the driver's attention from the road and thereby jeopardize safety.

Until recently, the driver-vehicle interface has evolved incrementally, with small successive changes in interaction and display features, so as to accommodate immediate needs as they arise. However, due to the increasing in-vehicle interface complexity, it is becoming clear that there is an imminent need for integration of these sub-systems in the car's driver interface into a coherent whole such that the driver can handle the vehicle safely and efficiently (Dingus & Hulse 1993).

In current vehicles, increasing amounts of information must be displayed to the driver (including the vehicle's on-the-road activity and performance, the vehicle's status and health, navigation, communication, and entertainment). This means that interaction and information presentations (e.g. displays) must be systematically designed. But since display space is always limited there is obvious competition for display space requiring a framework for classifying the cockpit components according to some basic principles related to tasks, urgency, timeliness, frequency and duration of use, passive or active interaction, and other operating criteria. The most crucial consideration is to make sure that the driver's attention to the road and the driving task is uncompromised by subsidiary activities.

Current user interfaces in cars consist of a main instrument cluster that includes all the vehicle and driving information, an upper and lower center stack that includes comfort control, multimedia, navigation, and communication, and sometimes even side-mirror displays that present information about adjacent-lane vehicles. Future cockpit systems are expected to include additional communication components (e.g. SMS, WhatsApp, and other social media elements), commercial and roadside information (e.g. location-based services), as well as information about vehicles, infrastructure, and other road users such as pedestrians and cyclists. Given this massive onslaught of information components, and the fact that the driver still bears responsibility for the control of the car (Level 1 and 2 automation, see National Highway Traffic Safety Administration 2013), effective and efficient information presentation and interaction has become an ever-increasing challenge.

In response to this challenge, Head Up Displays (HUD) are being considered and implemented as an additional display medium that has the unique advantage of being located in the drivers' line of sight, hence enabling them to pay attention to the road without turning their head. Clearly, head up displays have an inherent limitation as to the limited amount of information that can be displayed without excessive cluttering of the driving field of view. Thus the judicious utilization of this medium requires careful consideration of what gets displayed, how, and when.

9.1.1 Head up displays

The interest in HUD for automotive system has received a boost from the recent "wearable" trend, where head mounted devices such as the Google glass and a variety of augmented reality devices makes them attractive to customers. Users who will become accustomed to these devices will want to use them in driving; and as such, information organization considerations of HUDs may have to be imported to wearable devices.

Generally speaking, a head up display presents superimposed information on the visual scene. In some cases the information is actually projected on the windshield, at times on a glass screen by the windshield (Wood & Howells 2001). It is generally accepted that head up displays provide task information in a way that groups together informational elements that are compatible (e.g. speed limit, vehicle speed, ACC set speed) in display proximity (Wickens & Carswell 1986; Shmueli et al. this volume). The overall objective is to provide all the relevant information for the task (e.g. speed maintenance and monitoring, radio station switching and frequency selection) so as to (1) make increasing amounts of information and interaction behavior available and easily accessible, (2) minimize gaze retention period, and (3) reduce mental workload in performing a given task (He 2013).

Underlying the use of head up displays in the automotive setting is the concern for increasing gaze-off-the-road periods. Longer than two second gaze-off-the-road periods have been linked to traffic accidents (Ablaßmeier et al. 2007). Automotive human factors guidelines call for reduced distraction time by using head up displays and placing commonly used and highly critical displays close to the line of sight, yet keeping in mind the need for the driver to focus primarily on the "out-the-window" visual information required for the driving task itself (see Green et al. 1995, p. 22). Studies have shown that Gaze Retention Period (GRP), defined as eye fixation period plus eye movement period, is shorter when looking through a head up system versus head down displays (Ablaßmeier et al. 2007).

While head up display technology has promising applications, there are important issues to be examined as this display can obscure the road scene or distract the driver from paying attention to important objects in the scene (pedestrians, other vehicles, etc.). Here we assume at the outset the head up displays are given and our aim is to maximize their utility. However the fundamental question as to the potential harm of these display warrants further examination (see He 2013).

In principle, it makes sense to use head up displays in order to supply the driver with relevant driving information for operating the vehicle, event detection, maneuvering, navigating, monitoring of the vehicle and environment, and even auxiliary information (phone calls, climate control, and radio interaction). At the same time it is crucial for the head up display to provide this information in a succinct and consistent manner, and be non-obtrusive as regards the visual field and the driving task. To meet these requirements, a systematic approach to the problem of information allocation is required.

9.1.2 Objectives

This research is aimed at providing a systematic approach to the problem of information organization with special emphasis on allocation of informational elements. It begins by characterizing information elements to obtain a driver-centric perspective on the set of elements provided. The approach also prioritizes this information as a way to allocate them to the display components as a function of cost-benefit principles, which leads to recommendations for organizing information.

9.2 Characterization of vehicle information

To apply a systematic approach to the problem of organizing information for display presentation, we first need to consider all the elements of information at hand and characterize them by their attributes (Jacobs 2014, p. 518). To do this we take the driver's point of view with a special emphasis on the main activities of driving and monitoring information. This will provide us with sufficient operational data to classify the information elements according to a prioritized presentation scheme that optimizes user interaction efficiency, safety and satisfaction.

For each information element to be displayed, we specify a set of associated attributes that characterize its full range of operational and consequential feature. The display designer needs to assign a value or property to each information attribute. These attributes are listed and discussed in the following sub-sections below:
1. Activity (driving, navigation, monitoring, auxiliary).
2. Information type (state, status, warning, alert).
3. Urgency (high, medium, low).
4. Timeliness (high, medium, low).
5. Frequency of interaction (high, medium, low).
6. Duration of interaction (high, medium, low).
7. Importance (high, medium, low)
8. Response requirement (none, discretionary, obligatory, imperative, imminent)
9. Activation mode (persistent, intermittent, selectable).

9.2.1 Activity

We distinguish between four main driver activities – *driving, navigation, monitoring*, and *auxiliary* (cf. Ablaßmeier et al. 2007; Bedny & Meister 1997; Norman 2005). The driving activity is further divided into operating the vehicle (speed, lane keeping and stability control) and tactical maneuvering (passing other vehicles and obstacles). Navigation activities support the sequential and planned maneuvers such as exits from highway, lane changes, and turns. Monitoring refers to the act of observing and registering the displayed information, understanding its meaning and implications (comprehension), and finally constructing a men-

tal model of the situation over time (projection) as well as understanding consequences (Solomon 1993). Auxiliary information refers to all activities that are not
associated with driving and monitoring such as setting radios, climate control,
and communications.

9.2.2 Information Type

We make a distinction between several types of information: state, status, alerts
and warnings.

a. *System state.* The value of a continuous quantity (e.g. vehicle speed, engine
   RPM, ambient temperature display, time, fuel level).
b. *System status.* A discrete variable that usually has only two values (e.g. lights
   on/off, ACC on/off) for nominal events.
c. *Alerts.* The annunciation of unusual or unexpected events that do not necessarily require immediate response (e.g. indication of an upcoming gas station
   or an incoming phone call).
d. *Warnings.* The annunciation of unwanted situations, either current or imminent, that require an immediate awareness, attention and response.

Information type can be further categorized according to whether the information
is direct (such as speedometer, tachometer, fuel level, instantaneous fuel consumption, automation aids status, navigational information) or derived (average
fuel consumption, average speed, trip time).

9.2.3 Urgency

Urgency is the extent and speed with which the information element must be
attended to and resolved by the driver. We define four levels of urgency (*high,
medium, low, none*).

9.2.4 Timeliness

Timeliness reflects the fact that an information element is relevant only within a
limited time window. When the time window expires the information becomes irrelevant or invalid. These include, for example, an upcoming road service station,
distance-to-maneuver, change route suggestion. For simplicity we divide timeliness into four levels that express the urgency and importance to act (*high, medium,
low, none*).

9.2.5 Duration of interaction

The length of time of the expected user interaction. We define four levels of inter-action duration (*high, medium, low, none*).

9.2.6 Importance

We rank information importance based on its utility to the driver, with benefit weighed positively and cost weighed negatively. We distinguish four levels of im-portance (*high, medium, low, none*)

9.2.7 Frequency of use

Reflects the intensity of driver interaction with or monitoring of the information element (*high, medium, low, none*).

9.2.8 Type of user response required

This category details the response requirement, on part of the driver, concerning the displayed information:

a. No response requirement (e.g. ambient temperature, date).
b. Discretionary response (e.g. for air conditioner status, radio station, incoming phone call, SMS messages).
c. Obligatory response (e.g. change gear, seat belt alert, high beam, low tire pres-sure).
d. Imperative response (e.g. engine failure, flat tire, low oil pressure, lane devi-ation).
e. Imminent response (e.g. collision warning).

9.2.9 Activation mode

The last attribute, *activation mode*, refers to how the information is presented. This can either be presented constantly (persistent), presented only when trig-gered by the system or environment (intermittent), or when selected by the driver for display (selectable).

Tab. 9.1: Information attributes.

	Activity	Information type	Urgency	Timeliness	Criticallity (perception of risk)	Frequency of user access	Duration of inter-action	Response type	Activation mode
Odometer	auxilary	state	none	none	low	low	low	none	persistent
Trip odometer	navigating	state	none	none	low	low	low	none	persistent
Door ajar icon	monitoring	warning	high	none	high	none	none	imperative	intermittent
High engine-temperature icon	monitoring	warning	high	low	high	none	none	imperative	intermittent
Engine-temperature gauge	monitoring	state	none	none	medium	low	low	none	intermittent
Low tire pressure icon	monitoring	warning	medium	none	high	none	low	obligatory	intermittent
Fuel level gauge	monitoring	state	none	none	high	none	none	obligatory	intermittent
Fuel low indicator	monitoring	warning	medium	low	high	none	none	obligatory	intermittent
Fuel range	navigating	annunciation	low	low	medium	low	low	none	persistent
Turn signal	driving	alert	none	none	medium	high	low	none	intermittent
PRNDL indicator	driving	status	none	none	medium	medium	low	none	persistent
Gear shift advisory (up, down)	driving	alert	medium	none	medium	medium	low	discretionary	intermittent
Speedometer	driving	state	none	none	high	high	low	none	persistent
Speed limit (traffic sign, etc.)	driving	alert	none	none	high	high	low	discretionary	intermittent
Speed limit indicator warning	driving	warning	high	none	medium	none	none	discretionary	intermittent
ACC on/off	driving	status	none	none	low	medium	low	none	selectable
ACC set speed	driving	state	none	none	high	medium	low	none	selectable
ACC engaged call	driving	status	none	none	high	medium	low	none	selectable
ACC set gap	driving	state	none	none	medium	medium	low	none	selectable
Lane centering engaged	driving	status	none	none	high	medium	low	none	selectable
Side blind zone	driving	alert	high	none	high	none	none	obligatory	intermittent
Lane Departure Warning (LDW)	driving	alert	high	high	high	none	none	obligatory	intermittent
Traffic events (e.g. construction)	navigating	alert	low	medium	medium	none	none	discretionary	intermittent
Vehicle ahead indicator	driving	alert	low	none	medium	none	none	none	intermittent
Compass	navigating	state	none	none	low	none	low	none	persistent
Tachometer	driving	state	none	none	low	medium	low	none	persistent
Turn-by-turn indicator Expected	navigating	alert	medium	high	high	high	low	discretionary	intermittent
Time of Arrival (ETA)	navigating	state	none	none	low	low	low	none	persistent
Map and vehicle location	navigating	state	none	none	high	high	medium	none	selectable

Tab. 9.1: (continued)

Activity	Information type	Urgency	Timeliness	Criticallity (perception of risk)	Frequency of user access	Duration of inter-action	Response type	Activation mode
Hazard warning flasher monitoring	alert	none	none	high	medium	low	none	selectable
Electronic Stability Control (ESC) monitoring	status	none	none	medium	low	low	none	selectable
Traction control monitoring	status	none	none	medium	low	low	none	selectable
Clock monitoring	state	none	none	high	medium	low	none	persistent
External temp monitoring	state	none	none	low	low	low	none	persistent
Oil pressure warning monitoring	warning	high	none	high	none	none	urgent	intermittent
Light bulb warning monitoring	alert	high	none	high	none	none	obligatory	intermittent
Front fog lights monitoring	status	none	none	medium	low	low	none	selectable
Rear fog lights monitoring	status	none	none	medium	low	low	none	selectable
Headlights low monitoring	status	none	none	medium	medium	low	none	selectable
Headlights high monitoring	alert	none	none	medium	low	low	none	selectable
Parking lights monitoring	status	none	none	medium	low	low	none	selectable
Parking brake on while driving monitoring	alert	high	none	high	none	low	imperative	intermittent
Forward collision alert monitoring	warning	high	none	high	none	none	imminent	intermittent
Safety belt status indicator(s) monitoring	status	high	none	high	low	low	discretionary	intermittent
Side passenger Air Bag Status monitoring	status	none	none	medium	low	low	none	persistent
Air bag malfunction monitoring	warning	high	none	high	none	none	imperative	intermittent
Elec. charge system light monitoring	alert	high	none	high	none	none	obligatory	intermittent
Check engine light Brake monitoring	alert	medium	none	high	none	none	discretionary	intermittent
System warning ABS monitoring	warning	high	none	high	none	none	urgent	intermittent
Brake warning light monitoring	warning	high	none	high	none	none	urgent	intermittent
Battery voltage warning monitoring	warning	high	none	high	none	none	obligatory	intermittent
Radio setting auxiliary	status	low	none	high	low	high	none	selectable
HVAC setting auxiliary	status	medium	none	high	low	medium	none	selectable
Incoming phone call indicator auxiliary	alert	high	high	medium	none	low	discretionary	intermittent
Outgoing phone call auxiliary	status	none	none	medium	medium	high	none	selectable

9.3 Allocation of information

This set of nine attributes constitutes the basis on which a methodology can be developed for the organization of displayed information. Table 9.1 shows the implementation of these nine attributes with respect to a sample list of commonly displayed information elements. Each information element can be displayed on one (and sometimes more than one) of the available display locations in the cockpit: the *main instrument cluster* (left, center, right), the *upper center stack*, the *lower center stack* and/or the emerging *Head Up Display* (see next section). Each of the display locations has a limited capacity for information content and hence a decision mechanism for organizing the information on the display elements must be provided. A methodology for deciding how to organize the displayed information.

We first consider the display locations. Each display location is located at a different viewing angle from the main line of sight (that is, the driving direction), and at a different distance from the driver. Hence the interaction with each display distracts the driver to a degree that corresponds to its angle and distance from line of sight. The head up display has the advantage of minimal distraction from the driving task since its viewing angle coincides with the driving line of sight. On the other hand, interaction with the center stack causes significant distraction as its information is displayed well below the main line of sight and at a significant distance from the driver.

We can associate a "cost factor" with every display location that expresses its distractive effect based on the deviation from the driving line of sight and distance (the latter particularly relevant when manual interaction with the display surface is required). Additionally, the interaction with each information element takes a specific duration, cognitive effort, etc. Also, the interaction with each information element has its own characteristic frequency of occurrence (or intensity). Thus interaction (duration and intensity) costs are associated with the various information elements.

Next we consider the *utility*, or benefit, that the driver derives from interacting with the information when operating the vehicle. This utility can be evaluated from the table of attributes associated with the information elements discussed earlier. First, the designer can rank (or assign values to) the attributes. For example, the designer can rank the *activities* by defining an ordered preference list such as driving first, navigating second, monitoring third, and auxiliary-application fourth. The designer may also want to prioritize the list in terms of *information type* (warning first, alert second, state third, and status fourth). The prioritization can be further refined by assigning numerical weights to the components thereby enabling an formal optimization approach for solving the display organization problem.

To summarize, the interaction with each information element entails both a utility and a cost. The utility is derived from the value of the information to the driver. The cost is derived from the distraction that the interaction entails (measured by the deflection from line of sight and length or interaction as well as the cognitive load associated with the interaction), such as searching for a particular radio station.

As such, the problem of allocating information elements to display components can thus be cast as a classical resource allocation problem. Every allocation of information elements is associated with a utility value. This utility is obtained from the utilities and costs of the information attributes and the distractive cost of the display component. An allocation should thus be made so that the overall utility is optimized (i.e. the cost is minimized and utility is maximized).

The optimal allocation obtained from the above analysis must undergo a further review and possible adjustment. This is due to the fact that there is no *a priori* guarantee that the optimization alone will cause the informational elements to be clustered into coherent units. This is a separate issue because the clustering of information has independent value that cannot be computed based on the individual information elements alone. For example, we may decide to leave out an element because of its low prioritization, but include it anyway so as to not prevent a cluster from being coherent. The clustering problem has its roots in understanding the (strong/weak) interrelationships among informational elements and the opportunity to combine elements with strong relations into a coherent unit (Alexander 2002a, 2002b; Degani 2014). As such, a second iteration devoted to clustering analysis is recommended to complete the information organization methodology. The problem of organizing informational elements in clusters according to importance, color coding, symmetry and more using a Gestalt approach (Wertheimer 1923) was investigated by Shmueli et al. (this volume) and a formal methodology is provided there.

9.4 Head up display (HUD) and its information organization

We consider the role of the head up display to be primarily a safety-oriented feature aimed to support the activities of driving, navigating and monitoring. However, since drivers are increasingly engaged in auxiliary activities such as entertainment, convenience, and communication (that are *not* driving related) and these activities appear to be an inevitable and indispensable part of driving, the driver should be supported with regard to these activities as well, as part of the safety role of the head up display.

The secondary role of the head up display is convenience. Current vehicle cockpit designs show a trend towards placing vehicle controls (adaptive cruise control engagements) as well as the many entertainment and communication controls (volume, phone on/off) within easy reach of the driver (e.g. on the steering wheel). While currently the display of these interaction controls is usually scattered around the cockpit, an emerging trend is to display feedback for these important interactions in the head up display, thereby creating a composite steering-wheel-to-HUD interface system.

In view of the dual role of the head up display for providing both safety and convenience, its design must meet the following objectives: (1) to provide the driver with easily accessible information for driving, navigating, and monitoring tasks (as well as auxiliary and convenience activities) within a single display that is collocated with the main driving line of sight, thereby minimizing the need to divert attention to other interface components; (2) to provide the driver with a concise view of the current vehicle's driving state, system configuration, and the state of the operational environment (V2I, V2V, V2X). This concise view serves to enhance the driver's situation awareness about the vehicle, its driving conditions, and its road scene – thus enabling a *one-shot* understanding of what is going on.

Beyond role and objectives of the head up display in supporting driving and auxiliary activities, what remains unclear is which information elements should be presented and how to organize and display this information, given the head up display's limited display space. Furthermore, since the head up display is constantly present in the viewing field of the driver, the information must be selected and displayed in a nonintrusive manner so as to minimize interference with the driving scene.

The use of a head up display nevertheless has several additional (implicit) design challenges. The first is that the information must be *complete* in the sense that driver's dependency on other display components for routine activities is minimized (or even eliminated). Second, the information must be confined to driving and interaction-related activities. These challenges place a major burden on the design in view of its limited display space and requirement of minimal clutter. Third, the information displayed on the head up display has significant bearing on the design of other display components if a consistent and coherent human-machine system is to be achieved. For example, it may be desirable to display alerts and warning only by category – thereby requiring the driver to obtain additional detailed information about their specific nature elsewhere on the interface. This suggests that other display components need to be augmented with information resources and details (e.g. vehicle diagnostics).

Thus we need to address two main design issues: (1) deciding which information needs to be extracted to design a complete and concise head up display; and

(2) deciding how to organize the information so as to maximize situation awareness, minimize clutter and facilitate easy interaction:

9.4.1 Information completeness and conciseness

Based on our viewpoint that the head up display is primarily a safety-oriented display, we suggest a standardized structure for presenting information, which includes four levels. First and foremost is the actual driving activity that consists of vehicle operation and immediate maneuvering as well as object/event detection and response. Next there is the need to respond to navigation demands. Monitoring of vehicle systems (high beam, blinkers, cruise control ON), alerts and warnings (malfunctions), and information about operational environment (road signs, hazards, traffic, etc.) comprise the third level. Finally, the HUD also serves to support auxiliary activities such as entertainment, communication, comfort, and the like.

This standardized structure can be visualized as four concentric circles, where the most critical role (driving) is in the center (see Fig. 9.1), navigation is next, etc. This conceptual structure of information is aimed at meeting head up display objectives; namely safety and driver convenience, minimizing distraction and supporting situation-awareness/understanding.

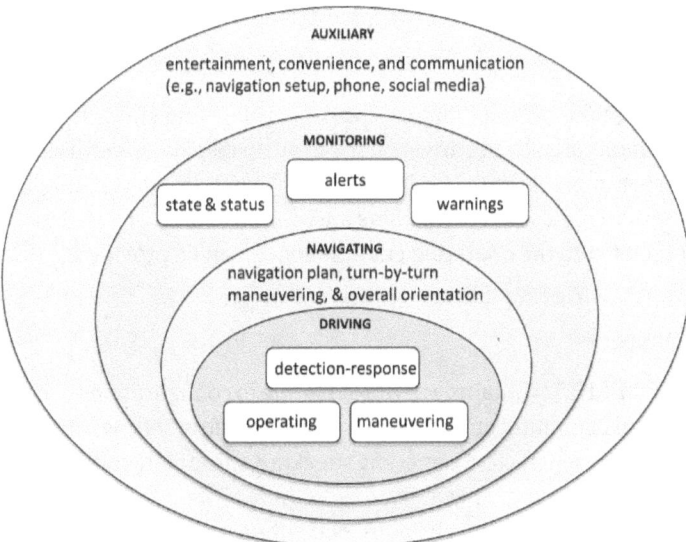

Fig. 9.1: Proposed structure of HUD information.

Driving

The driving level has three components: The *operating* component includes all the information related to speed, safe distance, lane control, etc. (presumably this information must always be present while the vehicle is being driven). The second component is the *maneuvering* that displays information concerning current, upcoming, and perhaps even future maneuver(s) such as turns, lane changes, and highway exists, etc., along with the corresponding parameters such as distance and/or time to maneuver. The set of maneuvers displayed on the head up display are related to both tactical maneuvers (dealing with road hazards and passing a vehicle, e.g.) as well as navigation related maneuvers. The operating and maneuvering components, in addition to *object and event detection and response* (the third component), can be linked visually to generate a functional unit for driving.

Navigation

The navigation level of the head up display provides the driver with information about navigation maneuvers (which are part of the overall navigation plan), parameters associated with these maneuvers (distance to, time to, etc.). Another important function of the navigation presentation is to provide the driver with *orientation* information. The key questions when making design decisions have to do with the role of safety in the navigation information provided on the display and whether it contributes to driver convenience.

Monitoring

The monitoring level of the head up display consists of state and status information that need to be monitored by the driver on the head up display as well as information that the driver interacts with (such as headway gap settings and adaptive cruise control modes). This is also where *alerts* and *warning* – some related to the vehicle and other related to the operating environment – can be provided.

Auxiliary

The auxiliary unit of the HUD contains activities relating to configurations of comfort, convenience, and communication features (climate control setting, infotainment, and phone, social media, etc.). Generally speaking, these activities are dealt with one at a time and are interchangeable.

The point is that a standardized structure can support designers in making sure that informational elements that have been selected for the head up display are not placed in disarray. From there on, the designer's challenge is to combine the utilization of the display "real estate" with the standardized structure to contain the information. Some clues to how this is done can be obtained from the human factors literature on configural displays as a technique for integration (Bennett & Flach 2011; Vicente 2002; and Shmueli et al. this volume).

9.5 Principles of HUD information organization

In this section we provide some general principles for HUD display organization. Some of these principles are based on the above discussion while others are based on information organization principles observed in and deduced from existing aviation head up displays where this technology has reached a certain level of maturation (see Wood & Howells 2001; Weintraub & Ensing 1992).

1. The head up display is partitioned into functional units whose placement is generally fixed. The display should be partitioned in a consistent manner such that relative feature locations are preserved (e.g. driving information is always below warnings, auxiliary features are always located at the bottom of the display).
2. Information should be presented in a manner that does not compete or interfere with the visual scene. As such, the information presentation should have minimal cluttering. Images and features need to be thin-lined, sharp and terse.
3. The head up display presentation structure should be somewhat consistent with the display structure and content on the cluster and center stack (e.g. head down displays).
4. The informational content on the display must be selected to minimize the dependency of the driver on monitoring other sources of information.
5. The representation on the head up display should be aligned with the line of sight (and direction of driving).
6. Graphics should be featured only to the extent that they convey concrete information and not for aesthetics purpose.
7. Color-coding is important for head up display presentation and a consistent color-coding is advised.
8. Head up displays have different display modes (user selected or automatically triggered) and may change dynamically given a new situation, task, or driver state. The transitions and display transformation necessary for switching between these modes can be supported by application of guidelines 1, 2, 3, and 4.

9.6 Review of existing Head Up Displays (HUDs)

Below we review and critique four representative head up displays. All these displays present primarily "driving" and "navigation" information with occasional alerts and warnings. Each review begins with a short description of the display and then lists the informational elements seen in the figures. We then discuss some of the design aspects of these displays in light of our guiding principles.

9.6.1 "Sporty" head up display

There are several display modes for this system. The one in Fig. 9.2 is the "sport" mode. Here the RPM gauge is the centerpiece of the display. The digital speed is integrated into the analog RPM. The gear position is placed to the right of the RPM and the distance-to-maneuver indicator at the far right. The warnings and alerts (here oil pressure low) are in the upper right corner.

Fig. 9.2: "Sporty" head up display.

Informational elements
- Tachometer (analog)
- Vehicle speed (digital)
- G-force meter (digital)
- Gear position
- Oil pressure warning (vertical)

The display is partitioned into fixed functional units in a consistent manner (relative feature location is preserved). The design of this display appears to be based on the perception that the driver is vehicle-performance oriented. Functionally there is no need to know the G-force and the RPM is justified solely based on the performance-orientation assumption. The fact that this head up display draws major attention and interferes with the road scene seems to be ignored, and is aggravated by the heavy analog graphics.

9.6.2 Simplistic HUD

The apparent aim here is to provide a simplistic presentation (see Fig. 9.3). In the center line of sight is the speed (digital) and road signs (school crossing sign). On the left is an icon for side blind zone indication (active safety) and on the right is navigation information

Fig. 9.3: Simplistic HUD.

Informational elements
– Vehicle speed (digital)
– Traffic signs (graphic)
– Side blind zone (icon)
– Maneuver ahead (schematic)
– Distance to maneuver

The head up display is partitioned into functional units in a consistent manner with preservation of relative feature location. The head up display information does not compete or interfere with the visual scene in that the design is very strict with right angles and minimal cluttering. Images and features are thin-lined, sharp and terse. The geometric representation is aligned with the line of sight and conveys concrete information. The overall organization of the display is a side-by-side. Finally, there seems to be little effort to integrate informational elements. It is not quite clear how this design could accommodate more advanced (automated, e.g.) driving features.

9.6.3 Colorful head up display

This is a very colorful head up display (see Fig. 9.4). It has three main elements indicating speed limit, maneuver ahead, distance to maneuver (or gap setting), and a horizontal speed bar that unites them all.

Fig. 9.4: Colorful HUD.

Informational elements
–   Speed limit (digital)
–   Maneuver ahead (schematic)
–   Distance to maneuver
–   Vehicle speed (horizontal bar)

The display competes with the visual scene due to the saturated use of colors. Images and features are heavy-lined and cause clutter. The (horizontal) analog speedometer is misaligned with the direction of driving. Generally speaking, there is limited systematic organization of functional units.

9.6.4 Graphically-rich head up display

This is a very intense and graphically rich display presentation (Fig. 9.5). It has a graphic rich presentation of the upcoming maneuver superimposed on the road. To the left there is the usual vehicle speed and adaptive cruise control and lane keeping assist presentation in a pyramid style.

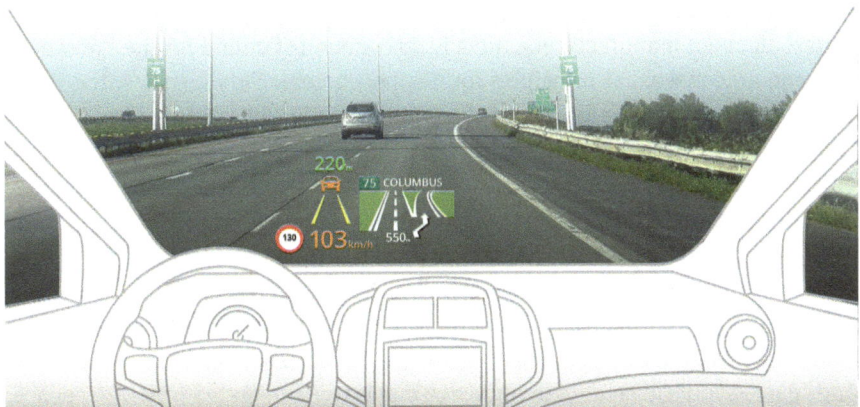

Fig. 9.5: Head up display with rich graphics.

Informational elements
- Speed limit (icon)
- Vehicle speed (digital)
- Lane keeping assist
- Vehicle ahead (icon)
- Distance to maneuver
- Maneuver ahead (pictorial)

This display is structurally similar to the display presented in Fig. 9.3. The driving component, which is schematic, is complemented by a navigation component that attempts to be realistic, thereby simulating augmented reality. The richness of the projected information competes dramatically with the road scene and may lead to confusion. There is aggressive use of graphics and colors but little information is conveyed.

9.7 Conclusion

The increased pace of growth in vehicle cockpit complexity and the plethora of information content competing for the driver's attention calls for a methodological, systematic, and rigorous approach to the design of information presentation (see Card, Mackinlay & Shneiderman 1999 for an inspirational call). This is true for all cockpit displays but is even more acute for head up displays that have limited available "real estate" if the cost of occlusion of the road scene is to be kept at bay. Although such displays are being offered in an increasing number of vehicles by almost all major manufacturers, there is no common standard for their organization and no consensual principles for display content. This suggests that in the future, a multitude of information elements may migrate into head up displays, creating clutter, occlusion, and distraction (see Fig. 9.6 which is an artistic "spoof" on the topic).

The problem of how to organize information in an integrated and coherent way is not new (Degani et al. 2009). The problem has its roots in art, where organization of the visual space, as a way to reflect an internal psychological space, has been explored by artists throughout the centuries of human development (Jung 1955/1972). When builders and later architects began to worry about floor- and space-layouts to create a living space that is well integrated with other spaces and is functionally workable, they employed similar principles of organization (Alexander 2002a) – which commonly are concerned with interrelationships between elements (rooms, entrances, kitchen, and even groups of buildings and layout of piazzas and town squares) and the creation of attractive spatial patterns (Necipoglu 1995). Designing interfaces for human interaction with machines, computers, and mobile devices also focuses on the problem of how to best organize the information so that the (interaction) space is well integrated, accessible, and coherent (Barshi et al. 2012; Monmonier 1996) – but with a slight twist: In informational devices the interaction space is abstract and is constructed to allows for simplicity and relevancy of information – two concepts that are critical for devices that superimpose information on the line of sight such as head up displays and wearables (e.g. Google glass and augmented reality device).

As automotive head up displays and similar wearable devices become more reliable and affordable, there is an opportunity to use the display not just as an addition to the information elements that exist in head down displays (e.g. cluster,

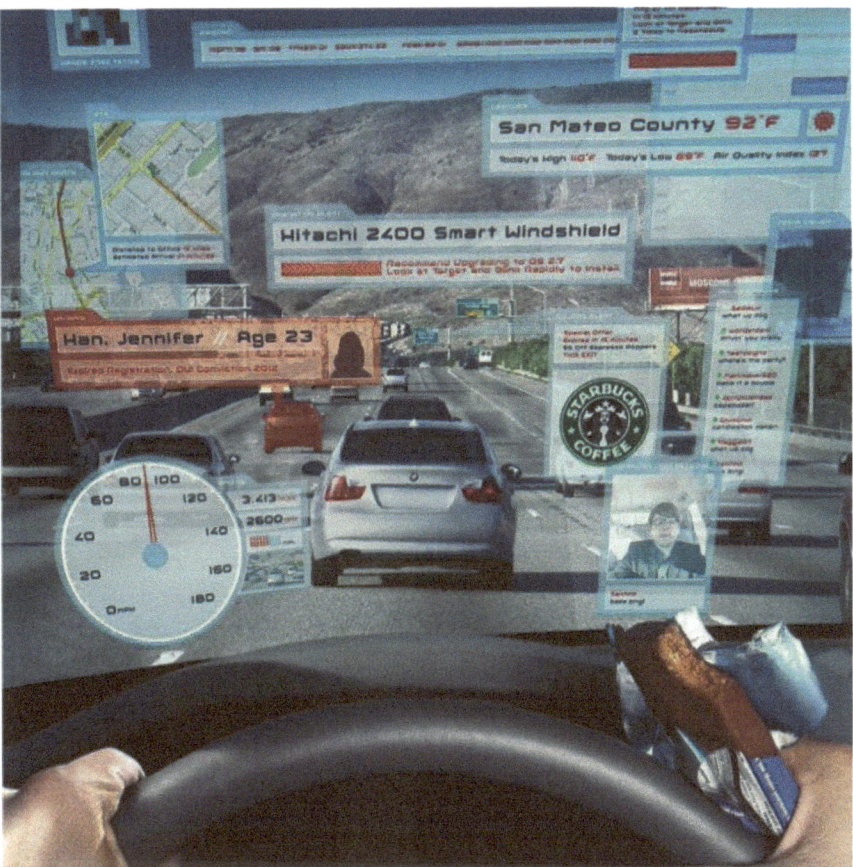

Fig. 9.6: The "Future" of HUDs (Photo and art: Erik Pawassar, with permission).

center stack etc.), but rather as a new instrument that can play a prominent role in the cockpit. Thus, we should begin to view the head up display as a separate display that deserves concern just like the other displays – and perhaps even more so given its benefits of being in the line of sight of the driver. Nevertheless the cost of distraction, clutter, and occlusion of the road scene requires special attention from the design community, not only because of its unique location in the cockpit but also because of its "special needs." The graphics of the head up display are different from the ones in head down displays and will be intrinsically different in terms of color-coding, background, backlighting, and stroke width. At the same time, consistency and similarity of appearance with identical and other informational elements on the head down displays is needed and must be maintained (cf. Tretten 2011).

For a century of automobile use, driver-vehicle interaction features and dashboard displays have converged to a fairly common and standardized configuration. Thus, while interaction style and control/display locations differed somewhat between manufacturers, standardization and generational consistency held true. This situation has begun to change with in-vehicle implementation of digital technologies and the rapid introduction of new advanced driver assistance systems, infotainment and communication features in current and upcoming cars. As a result of the sheer size of the information content and its novelty, information organization in modern vehicles is in a state of flux. This is an unprecedented opportunity to develop formal design methodologies for in-vehicle information presentation to improve current design practices and at the same time provide guidance for upcoming display technology such as wearable devices and enhanced- and augmented-reality displays.

Abbreviations

ADAS   Advanced Driver Assistance Systems
ACC   Adaptive Cruise Control
HUD   Head Up Display
LC   Lane Centering
GRP   Gaze Retention Period

References

Ablaßmeier, M, Poitschke, T, Wallhoff, F, Bengler, K & Rigoll, G 2007, 'Eye gaze studies comparing head-up and head-down displays in vehicles', *2007 IEEE International Conference on Multimedia and Expo*, pp. 2250–2252.

Alexander, C 2002a, *The phenomenon of life*, The Center for Environmental Structure, Berkeley, CA.

Alexander, C 2002b, *The process of creating life*, The Center for Environmental Structure, Berkeley, CA.

Barshi, I, Degani, A, Iverson, D & Lu, PJ 2012, 'Using medieval architecture as inspiration for display design: Parameter interrelationships and organizational structure', *Proceedings of the Human Factors and Ergonomics Society Annual Meeting*, vol. 56, pp. 1799–1803.

Bennet, KB & Flach, JM 2011, *Display and interface design: Subtle science, exact art*, CRC Press, Boca Raton, FL.

Broy, M, Krüger, I, Pretschner, A & Salzmann, C 2007, 'Engineering automotive software', *Proceedings of the IEEE*, vol. 95, no. 2, pp. 356–373.

Card, SK, Mackinlay, JD & Shneiderman, B 1999, *Information visualization: Using vision to think*, Morgan-Kaufmann, San Francisco.

Degani, A 2013, 'A tale of two maps: The story of the London underground "diagram"', *Ergonomics in Design*, vol. 21, no. 3, pp. 7–16.

Degani, A, Jorgensen, C, Shafto, M & Olson, L 2009, *On organization of information: Approach and early work*, NASA Technical Memorandum No. 215368, NASA Ames, Moffett Field, CA.

Green P, Levison W, Paelke G, & Serafin, C 1995, *Preliminary human factors design guidelines for driver information systems*, FHWA-RD-94-087, Federal Highway Administration, Washington, DC.

He, J 2013, 'Head-up display for pilots and drivers', *J Ergonomics* vol. 3, no. 3.

Jacobs, E 2004, 'Classification and categorization: A difference that makes a difference', *Library Trends*, vol. 52, no. 3, pp. 515–540.

Jung, CG 1955/1972, *Mandala symbolism*, Princeton University Press, Princeton, NJ.

Dingus, TA & Hulse, MC 1993, 'Some human factors design issues and recommendations for automobile navigation information systems', *Transportation Research, Part C, IC*, pp. 119–131.

Leen, G & Hefferman, I 2002, 'Expanding automotive electronic systems', *IEEE Computer*, vol. 35, no. 1, pp. 88–93.

Minoiu Enache, N, Netto, M, Mammar, S & Lusetti, B 2009, 'Driver steering assistance for lane departure avoidance', *Control Engineering Practice*, vol. 17, no. 6, pp. 642–651.

Monmonier, M 1996, *How to lie with maps*, University of Chicago Press, Chicago.

National Highway Traffic Safety Administration 2013, *Preliminary statement of policy on automated vehicles*, May 30,

Norman, D 2007, *The design of future things*, Basic Books, New York.

Necipoglu, G 1995, *The Topkapi Scroll – Geometry and ornament in Islamic architecture*, The Getty Center for the History of Art.

Richtel, M 2011, 'U.S. safety board urges cellphone ban for drivers', *New York Times*, December 13.

Rasmussen, J & Vicente, KJ 1989, 'Coping with human errors through system design: Implications for ecological interface design', *International Journal of Man-Machine Studies*, vol. 31, pp. 517–534.

Solomon, RC 1993, *The passions: Emotions and the meaning of life*, Indianapolis, IA.

Tretten, P 2011, *Information design solutions for automotive displays: Focus on HUD*. Unpublished PhD Thesis. Luleåtekniska universitet.

Vicente KJ 2002, 'Ecological interface design: progress and challenges', *Human Factors*, vol. 44, pp. 62–78.

Weintraub, DJ & Ensing, MJ 1992, *Human factors issues in head-up display design: The book of HUD*, Crew System Ergonomics Information Analysis Center, Wright-Patterson AFB, Ohio.

Wertheimer, M 1923, *Laws of organization in perceptual forms*. Published in 'A source book of Gestalt psychology', Routledge & Kegan Paul, London, pp. 71–88.

Wickens, C & Andre, A 1990, 'Proximity compatibility and information display: Effects of color, space, and abjectness on information integration', *Human Factors*, vol. 32, no. 1, pp. 61–77.

Wickens, C & Carswell, C 1995, 'The proximity compatibility principle: Its psychological foundations and its relevance to display', *Human Factors*, vol. 37, no. 3, pp. 473–494.

Wood, RB & Howells, PJ 2001, 'Head up displays', in *The Avionics Handbook*, ed RC Spitzer, CRC Press, Boca Raton, FL, Chapter 4.

Index

Zeitfracht Medien GmbH
Ferdinand-Jühlke-Straße 7
99095 Erfurt, Deutschland
produktsicherheit@kolibri360.de